THE HAND

Instructional Aids Committee
of the Continuing Education Committee
American Society for Surgery of the Hand

RICHARD I. BURTON, MD, *Chairman*
Rochester, New York

LOUI G. BAYNE, MD
Atlanta, Georgia

JAMES L. BECTON, MD
Augusta, Georgia

LEONARD T. FURLOW, Jr., MD
Gainesville, Florida

JAMES H. HOUSE, MD
Minneapolis, Minnesota

FRED B. KESSLER, MD
Chairman Educational Committee,
American Society for Surgery of the Hand,
Houston, Texas

GORDON B. McFARLAND, Jr., MD
Coordinator, Division of Education,
New Orleans, Louisiana

CHRIS P. TOUNTAS, MD
St. Paul, Minnesota

THE HAND

EXAMINATION AND DIAGNOSIS

AMERICAN SOCIETY FOR
SURGERY OF THE HAND

SECOND EDITION

CHURCHILL LIVINGSTONE
EDINBURGH LONDON MELBOURNE AND NEW YORK 1983

CHURCHILL LIVINGSTONE
Medical Division of Longman Group Limited

First Edition 1978
Second Edition 1983

ISBN 0 443 02310 7

British Library Cataloguing in Publication Data
American Society for Surgery of the Hand
The hand. —— 2nd. ed.
1. Hand —— Diseases —— Diagnosis
2. Hand —— Surgery
I. Title
617'.575075 RC951

Library of Congress Cataloging in Publication Data
Main entry under title:
The Hand, examination and diagnosis.
Includes index.
1. Hand —— Wounds and injuries. 2. Hand —— Abnormalities. 3. Hand —— Diseases. 4. Hand —— Examination. I. American Society for Surgery of the Hand. [DNLM: 1. Hand injuries —— Diagnosis. 2. Hand deformities, Acquired —— Diagnosis. 3. Hand —— Abnormalities. WE 830 H2308]
RD559.H359 617'.575075 82-4140
AACR2

Printed in U.S.A.

FOREWORD

Although this monograph outlines the examination of the hand, it is important to stress that the function of the hand begins within the central nervous system. Modern surgery of the hand is not a limited regional speciality. No medical or surgical specialist treats a greater variety of pathological entities than the hand surgeon.

The armamentarium of the hand surgeon includes the established procedures of the orthopaedic surgeon and the plastic surgeon, plus innovative techniques such as tendon implants for the severe injury or rheumatoid arthritis. The technical skills of the neurosurgeon and vascular surgeon are modified for microsurgery, such as the replantation of amputated digits. Pediatric studies are utilized by the hand surgeon in the medical and surgical management of the congenital upper extremity anomaly.

The basic discipline of the hand surgeon is anatomy, and this monograph emphasizes the anatomical principles for the diagnosis and treatment of the abnormal hand.

The Instructional Aids Committee of the Division of Education has developed this text with enthusiasm and prolonged effort. The individual members have met in repeated workshops based on personal experience and research. The result is a valuable handbook.

The personality and activities of the individual are often recorded in the appearance and dexterity of their hand. This monograph attempts to define, in a limited way, how one might preserve and reconstruct that individuality.

George E. Omer, M.D.
President,
American Society for Surgery of the Hand, 1979

PREFACE

The hand is composed of material for touch of great sensitivity and a system of exact machinery of great specialization and refinement — all in a most complex array and condensed into a unit weighing less than a pound. With this amazing tool, we implement the desires of the human brain, whether requiring the speed and precision of the fingering hand of a concert violinist or the brute power grasp needed to wield a sledge hammer.

Sir Charles Bell,[1] the leading British anatomist, physiologist, and neurologist of the early 19th century, was among the first to recognise the unique qualities of the human hand: '. . . it is in the human hand that we perceive the consummation of all perfection, as an instrument. This superiority consists in its combination of strength, with variety, extent, and rapidity of motion . . . and the sensibility, which adapt it for holding, pulling, spinning, weaving, and constructing; . . .' With the hands the laborer supports a family, the parent loves and cares for a baby, the musician plays a sonata, the blind 'read', and the deaf 'talk'.

This essential organ, the hand, is often crippled by injury, by disease, or by birth defects.

To address this human need, there has emerged in

the last three decades a special area of expertise — Surgery of the Hand.

One of the earliest pioneers in Surgery of the Hand was Allen B. Kanavel of Chicago (1874-1938). His main contributions were in the understanding and treatment of infections of the hand, and his book on this subject is a classic. His efforts were furthered by his associate, Sumner Koch (1888-1976).

During this same era, Arthur Steindler (1878-1959) of Iowa City developed the principles and many of the details of tendon and muscle transfers for the disabled upper extremity — concepts which are still utilized today.

Perhaps the most influential person in the history of Surgery of the Hand was Sterling Bunnell. In order to improve the treatment of the hands disabled in combat, during World War II, he was appointed as a special consultant to the Secretary of War 'to guide, integrate, and develop the special field of Hand Surgery'. From the hand centers which he developed sprang a core of surgeons dedicated to the principles and philosophies of Dr Bunnell.

From this small group of Dr Bunnell's disciples emerged a group of surgeons from general, orthopaedic, and plastic surgery who, recognizing the uniqueness of this specialized organ, organized the American Society for Surgery of the Hand. Through its influence, in turn, similar Hand Societies have been founded in most countries of Europe, Asia, and Central and South America. These efforts have culminated in the formation of the International Federation of Societies for Surgery of the Hand.

Today the horizons for Surgery of the Hand have further expanded to include arthritis, congenital deformities, and even the replanation of amputated parts. As

Sterling Bunnell aptly summarized,[2] 'To recondition these members successfully is difficult. Surgical reconstruction of the hand requires special careful technique. ... It is a composite problem requiring the correlation of various specialities — orthopaedic, plastic, and neurologic surgery — the knowledge of any one of which alone is inadequate for repairing the hand.... As the problem is composite, the surgeon must also be ... The surgeon must face the situation and equip himself to handle any and all of the tissues of a limb.'

The intent of this brief monograph is to introduce some of the basic anatomic principles upon which is based this new but exacting and essential discipline.

Richard I. Burton, M.D.

REFERENCES

1. Bell Sir Charles 1833 The Hand, Its Mechanism and Vital Endowments, as Evincing Design. Coney, Lea and Blanchard, Philadelphia.
2. Bunnell S 1944 Surgery of the Hand. Lippincott, Philadelphia.

ACKNOWLEDGEMENT

The germinal concept of this book began at a meeting of the Task Force on Continuing Education in 1974 attended by Drs George Omer, Fred Kessler, James Becton, Edward Nalebuff and myself. It fell to Dr Becton to produce a working outline and preliminary drawings of this production. Each chapter was then amplified and supervised by an individual member of the Instructional Aids Committee of the American Society for Surgery of the Hand. Dr Richard Burton, chairman of that committee, has been tireless in his efforts to produce an academically sound, yet practical approach to the physical diagnosis of the hand.

Though this has truly been a product of the entire committee, I am especially grateful to Dr Becton not only for his original outline, but for his continued efforts; Dr Burton for his skill and diplomacy in guiding a true committee effort; Dr Omer for his guidance and Dr Kessler, without whose spark, continued leadership, and editorial supervision this work might not have been produced at all.

Gordon B. McFarland, Jr., M.D.
Coordinator, Division of Post-Graduate Education,
American Society for Surgery of the Hand, 1978

TABLE OF CONTENTS

PART 1. EXAMINATION

Introduction 2

1. History and General Examination 3

2. Examination of Specific Systems 11

PART 2. COMMON CLINICAL PROBLEMS

3. Lacerations 49

4. Common Fractures and Dislocations of the Hand 53

5. Acquired Deformities 64

6. Congenital Anomalies 84

7. Tumors 89

8. Infection 93

APPENDICES

Appendix 1. Abbreviations 100

Appendix 2. Anatomy — Summary 103

Appendix 3. Clinical Assessment Recommendations 106

INDEX 113

PART 1

EXAMINATION

INTRODUCTION

This text is a core of information on the diagnostic history and physical examination of the normal, diseased or injured hand. A method for thorough, systematic evaluation of the hand is presented so that with practice the reader can develop a routine for accurate examination to achieve a specific diagnosis.

A brief introduction to specific conditions of the hand is given, followed by illustrations of the more common disorders. A limited description of certain lacerations, fractures, dislocations, and deformities is included.

Specific treatment of each diagnosis is *not* discussed. The reader is referred to the standard texts and the current literature of hand surgery for detailed descriptions of treatment methods.

1

HISTORY AND GENERAL EXAMINATION

HISTORY

Before examining the hands, a detailed history of the present problem should be obtained:

A. What are the patient's age, occupation, and pursuits?
   Which is the dominant hand?
   Has there been a previous hand impairment or injury?

B. *In trauma problems* the history should include the following specific information:
   1. *When* did the injury occur and how much time has elapsed since the injury?
   2. *Where* did the injury occur? Was it at work, home, or play? Under what conditions — was the environment clean or dirty?
   3. *How* did the injury happen? What was the exact mechanism of the injury? (This helps to evaluate the amount of crush, contamination, blood loss, and level of injury to gliding parts.) What was the exact posture of the hand at the time of injury?
   4. *What* previous treatment has been administered?

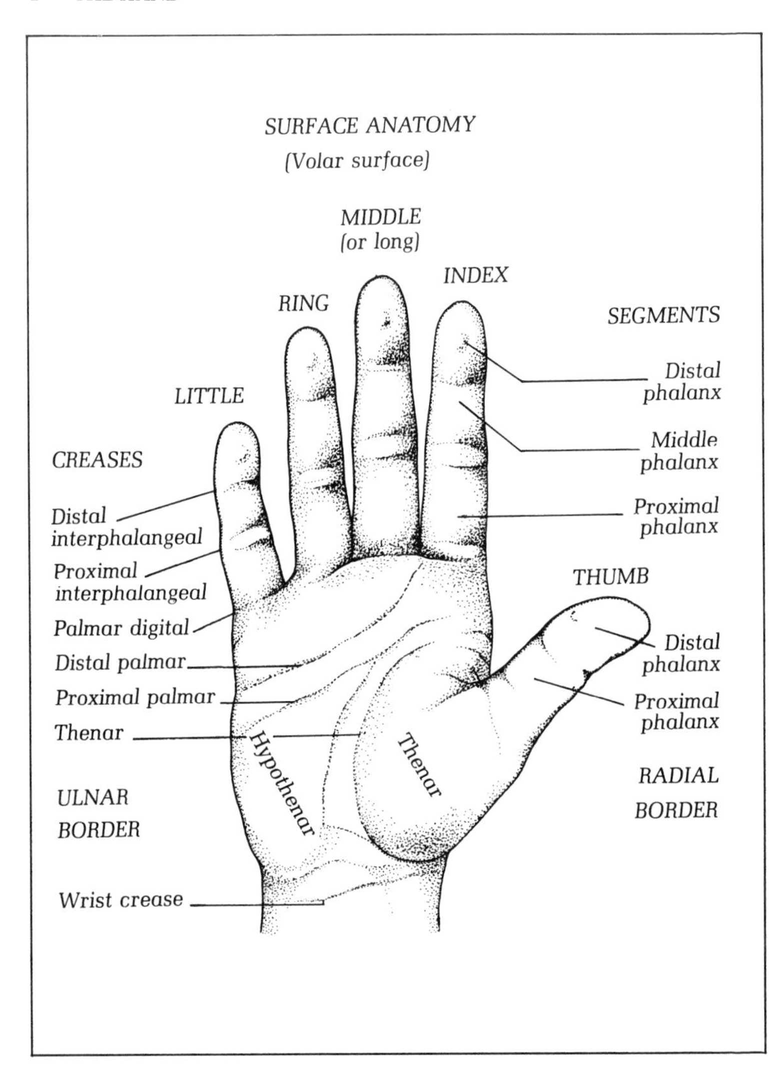

Figure 1
Surface anatomy of the hand

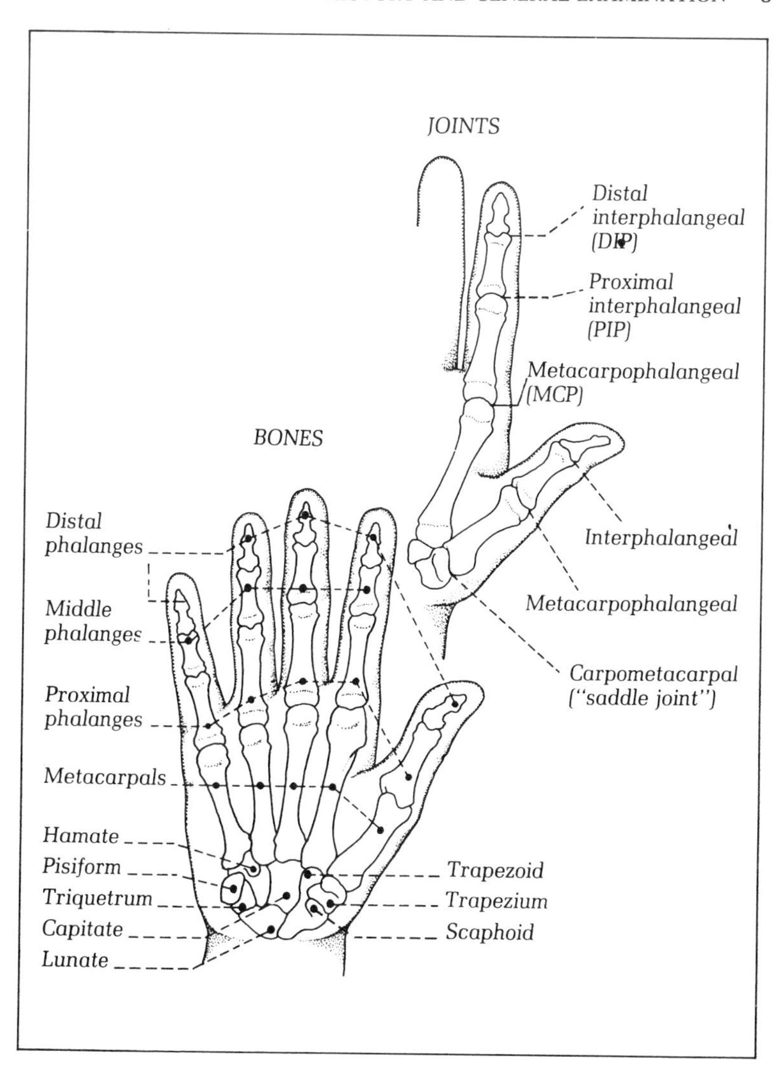

Figure 2
Skeleton of the hand and wrist

C. *In nontrauma problems* particular emphasis should be placed on:

1. *When* did the pain, sensory change, swelling, or contracture begin? In what sequence? Are these symptoms progressive?
2. How is *function* impaired in occupation, hobby, and activities of daily living?
3. Are *other joints or tendons* in this or other extremities painful in a similar way?
4. What *activities* make the pain worse?
5. At what *time* of day or night is the pain worse?

D. A review of the past medical history and a review of systems should be obtained as part of the complete evaluation of the hand.

TERMINOLOGY

In order to avoid confusion it is important that standard terminology for structures of the hand be used. The hand and digits have a dorsal surface, a volar or palmar surface, and radial and ulnar borders (Fig. 1). The palm is divided into the thenar, mid-palm, and hypothenar areas. The names of the digits are the thumb, index, middle (long), ring, and little fingers. The thenar mass or eminence is that muscular area on the palmar surface overlying the thumb metacarpal. The hypothenar is that muscle mass on the palmar surface overlying the little finger metacarpal. Each finger has three joints: the metacarpophalangeal (MCP), the proximal interphalangeal (PIP), and the distal interphalangeal (DIP) joints (Fig. 2). Note the location of the finger MCP joints in the palm near the distal palmar crease, with the palmar-digital creases and finger webs at the level of the middle third of the proximal phalanges.

The thumb has an MCP and only one interphalangeal (IP) joint. The carpometacarpal joint of the thumb is particularly important because of its mobility. There are proximal, middle, and distal phalanges in the fingers and only a proximal and a distal phalanx in the thumb.

The terminology used to describe the motion of the joints is illustrated in Figures 3A and 3B.

PHYSICAL EXAMINATION (see Ch. 2 for details)

The entire upper extremity should be exposed and evaluated when the hand is examined. Assessment of active shoulder motion, elbow motion, and pronation and supination of the forearm is essential. Motion of these joints is necessary for proper positioning of the hand for function. Any discrepancy between active and passive mobility should be noted.

When inspecting the hand, one should observe its color to assess circulation as well as the radial and ulnar pulses. The presence of swelling or edema should be noted as well as any abnormal posture or position. Skin moisture, localized tenderness, and sensibility must be evaluated.

After injury to the hand there is often secondary stiffness and limited range of motion of other joints of the extremity as well as the part involved. The range of both passive and active motion of the wrist, MCP joint, and IP joints of each digit should be measured and recorded. Grip and pinch strength should also be documented. The patient's ability to use the hand for simple function should be evaluated.

Accurate recording of the findings of the examination of the hand is most important. A simple

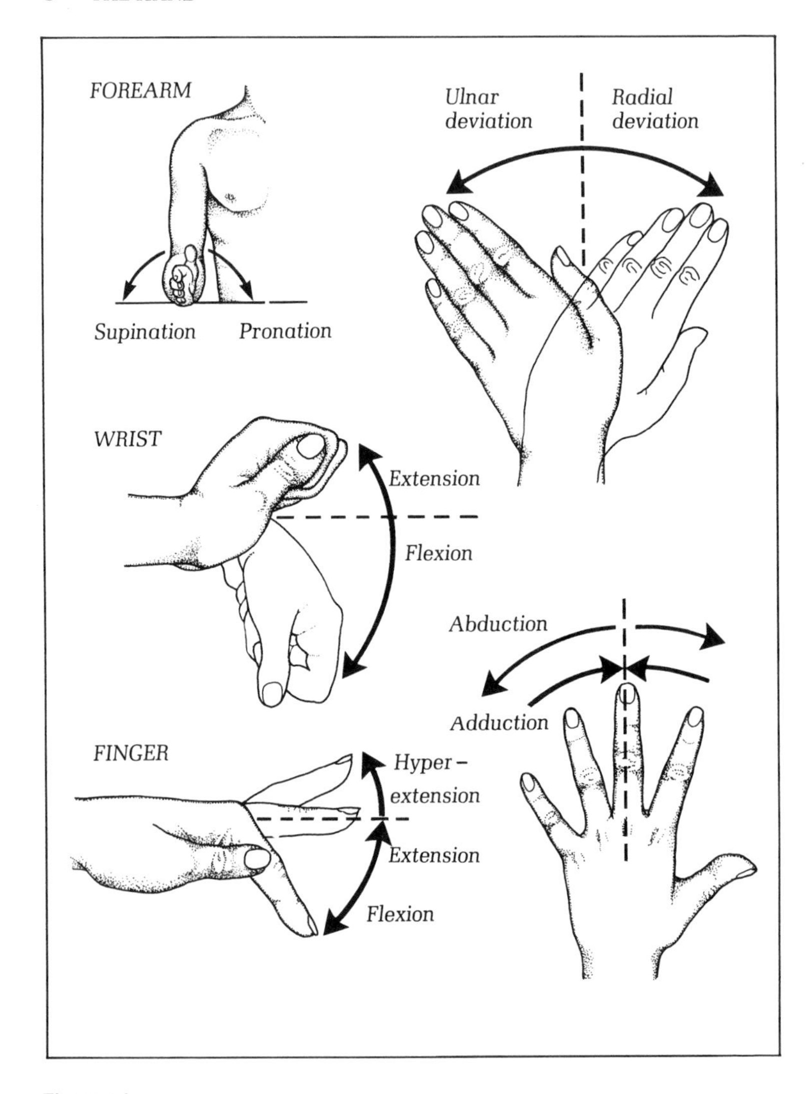

Figure 3A
Terminology of hand and digit motion

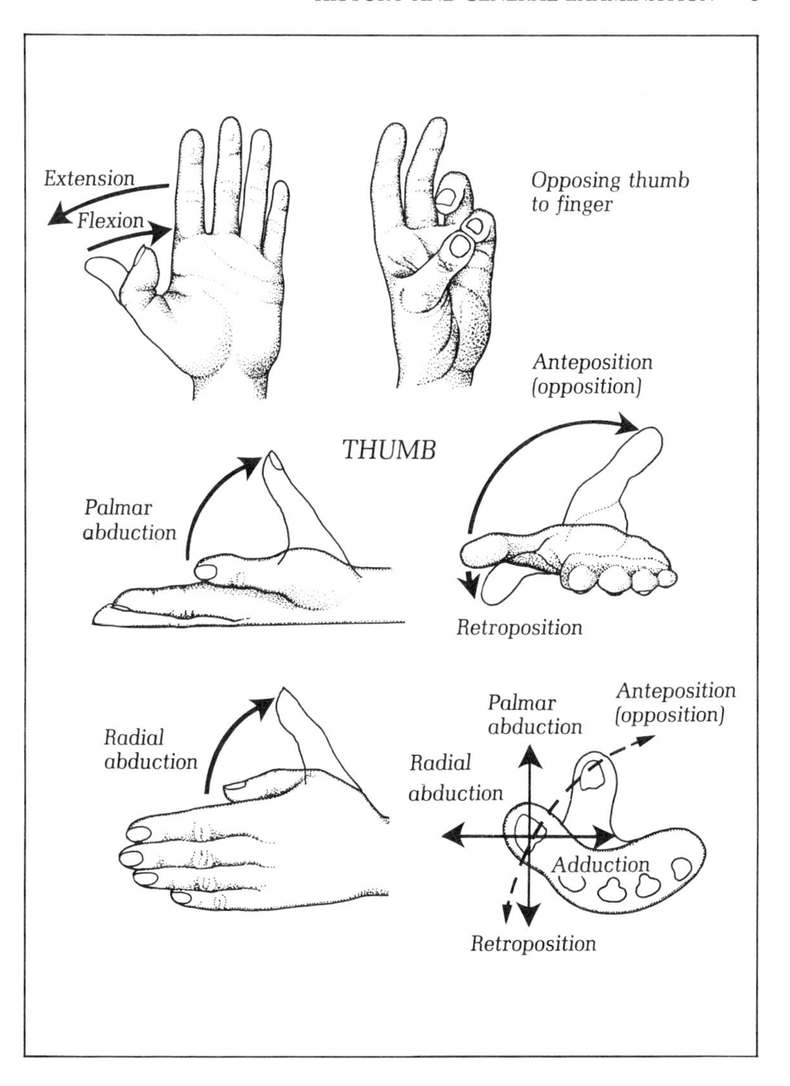

Figure 3B
Terminology of hand and digit motion

sketch of the hand with appropriate notations and measurements is often very helpful.

Subsequent re-examination of the hand is just as important as the initial examination and should be done each time the patient returns for follow-up. Only by making appropriate serial recordings during the following-up period can the examiner know for certain whether or not there is improvement in the patient's condition.

2

EXAMINATION OF SPECIFIC SYSTEMS

SKIN

The normal palmar skin is thick, tethered, irregularly surfaced, and moist, providing for traction and wearability. Normal skin on the dorsum of the hand is thin and mobile, permitting motion of the various joints. The dorsum of the hand is the common site of edema, which may limit flexion. The examiner should note the presence or absence of swelling, wrinkles, color, moisture, scars, skin lesions, and surface irregularities.

MUSCLES

The muscles that power the hand may be divided into extrinsic and intrinsic muscles. *The extrinsic muscles* have their muscle bellies in the forearm and their tendon insertions in the hand. They are further divided into extrinsic flexor and extensor muscles. The flexors are on the volar surface of the forearm and flex the wrist and digits, the extensors are on the dorsum of the forearm and extend the wrist and digits.

The intrinsic muscles have their origins and insertions within the hand.

These muscles should be systematically evaluated. Ask the patient to 'make a fist' and 'straighten out your fingers'; this gives the examiner a general idea of the active range of motion of the digits. However, it is necessary to examine each muscle group specifically.

Specific extrinsic muscle testing

Extrinsic flexor muscles

The function of the *flexor pollicis longus* (FPL) muscle, whose tendon inserts on the volar base of the distal phalanx of the thumb, can be evaluated by asking the patient to 'bend the tip of your thumb' (Fig. 4). The muscle strength is tested against resistance supplied by the examiner.

The flexor digitorum profundus (FDP) can be tested by asking the patient to 'bend the tip of your finger' (Fig. 5). The PIP joint is stabilized in extension by the examiner as the distal joint is actively flexed. As each finger is examined, the muscle is tested against resistance.

Each flexor digitorum superficialis (FDS) is individually tested by asking the patient to 'bend your finger at the middle joint' (Fig. 6). The other fingers must be stabilized in extension by the examiner so as to block profundus function. (The profundus tendons of the ulnar three digits share a common muscle belly, and thus independent flexion of any finger with the other digits restrained in extension requires intact FDS musculotendinous functions to that finger). The procedure is repeated for each finger.

The flexor carpi ulnaris (FCU), flexor carpi radialis (FCR), and *palmaris longus (PL)* are evaluated by asking

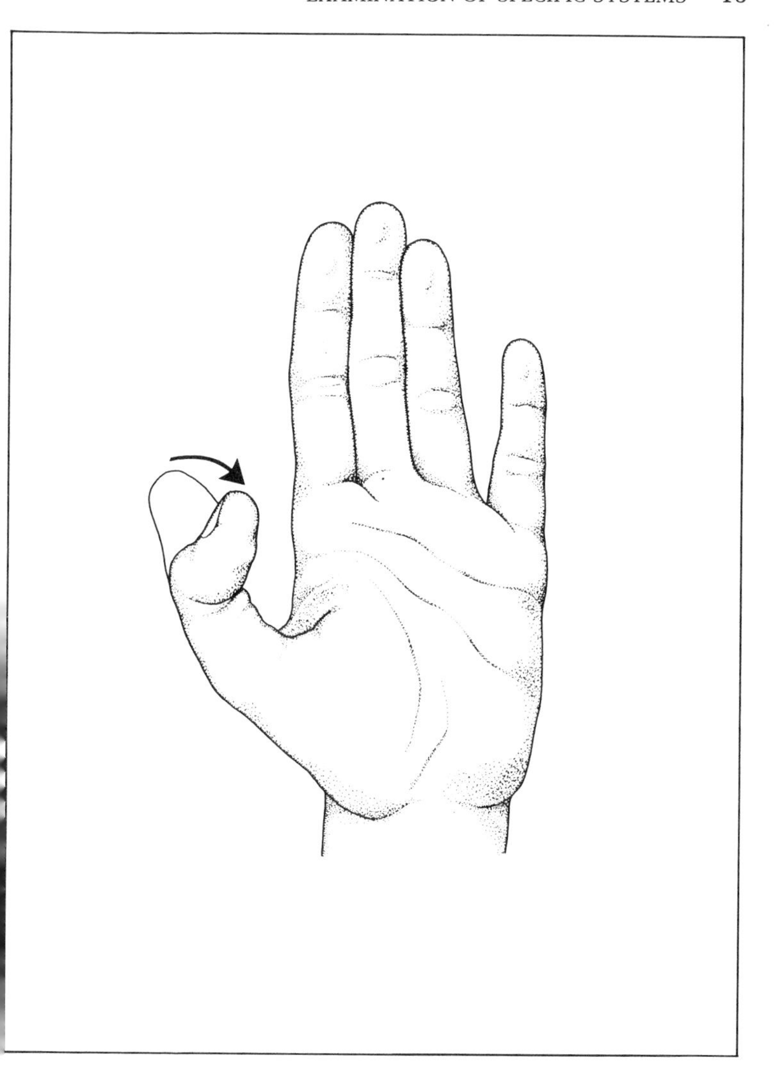

Figure 4
Testing for FPL musculotendinous function

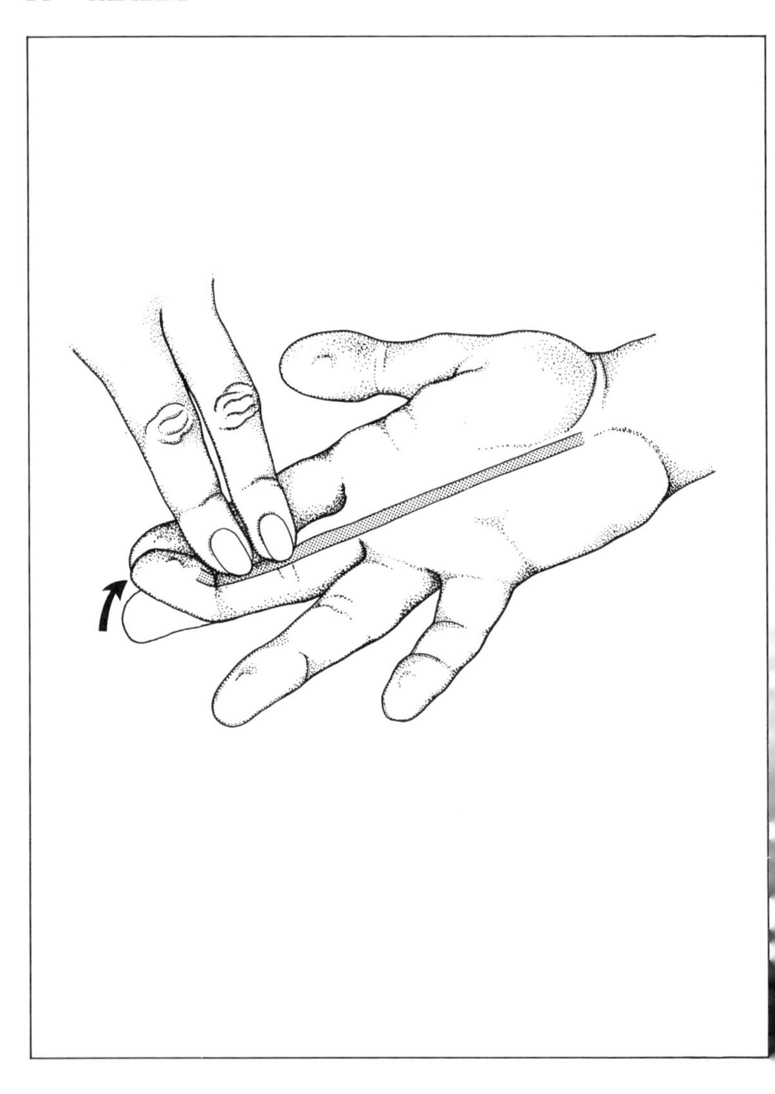

Figure 5
Testing for FDP musculotendinous function

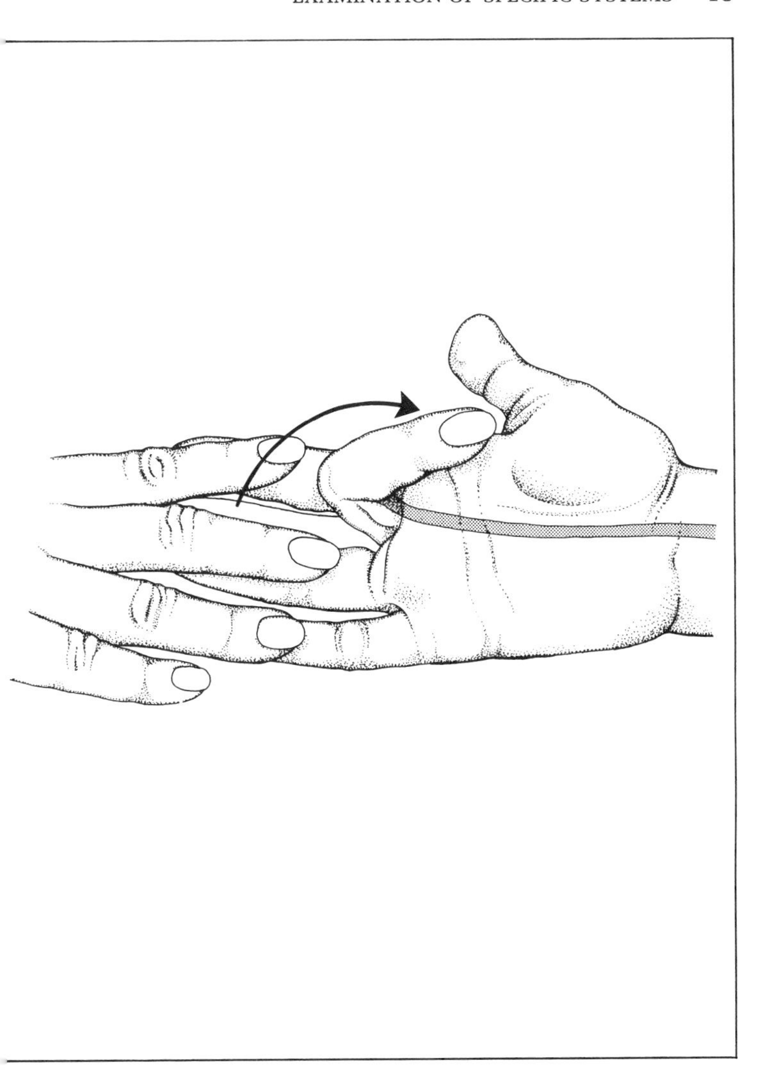

Figure 6
Testing for FDS musculotendinous function

the patient to flex his wrist while the examiner palpates the tendons of these muscles. The FCU inserts into the pisiform and the FCR into the volar aspect of the index metacarpal. The PL inserts into the palmar fascia. The PL will be noted to lie between the FCR radially and FCU ulnarly on the volar surface of the wrist during this manoeuvre, especially if the thumb is simultaneously opposed to the little finger.

Extrinsic extensor muscles

The extrinsic extensor muscle bellies of the hand overlie the dorsum of the forearm and their tendons pass over the dorsum of the wrist to insert in the hand (Fig. 7). They are arranged in six tendon compartments over the dorsum of the wrist. A systematic examination of the tendons passing through each compartment is done.

The first dorsal wrist compartment. The first dorsal compartment of the wrist contains the tendons of the *abductor pollicis longus* (APL), which inserts at the dorsal base of the thumb metacarpal, and the *extensor pollicis brevis* (EPB), which inserts at the dorsal base of the proximal phalanx of the thumb. These are evaluated by asking the patient to 'bring your thumb out to the side' (Fig. 8). The examiner can palpate the taut tendons over the radial side of the wrist going to the thumb.

The second dorsal wrist compartment. The second dorsal wrist compartment contains the tendons of the *extensor carpi radialis longus* (ECRL) and the *extensor carpi radialis brevis* (ECRB) muscles (Fig. 9). They insert at the dorsal base of the index and middle metacarpals,

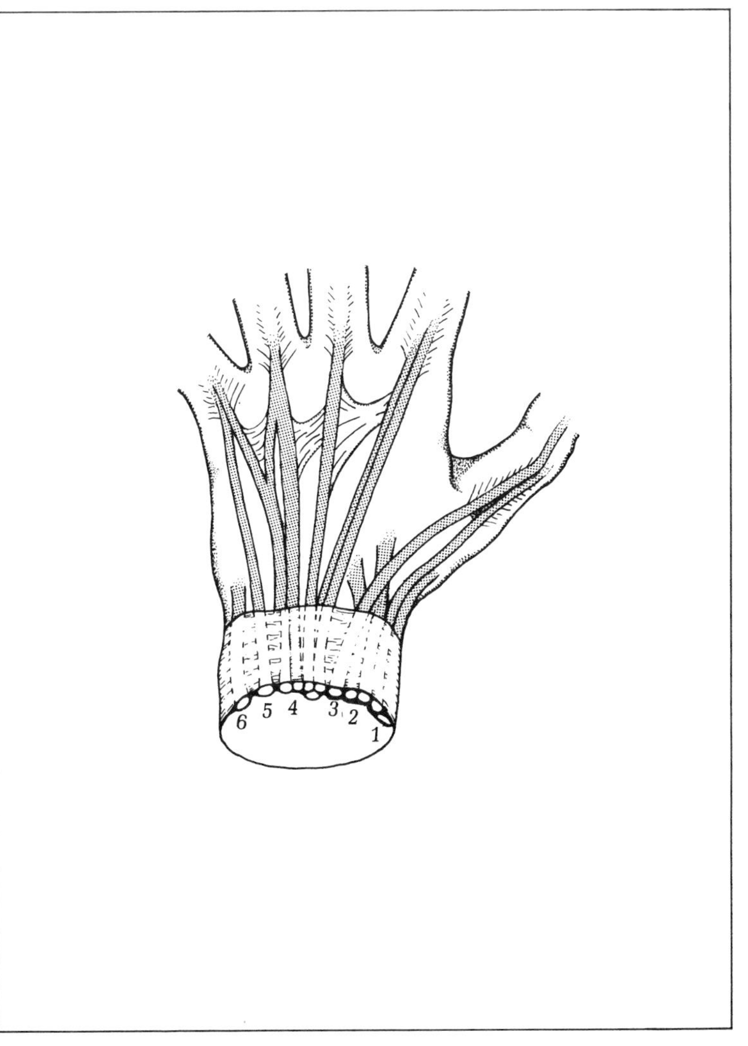

Figure 7
Arrangement of extensor tendons at the wrist into 6 compartments: dorsal and cross-section view

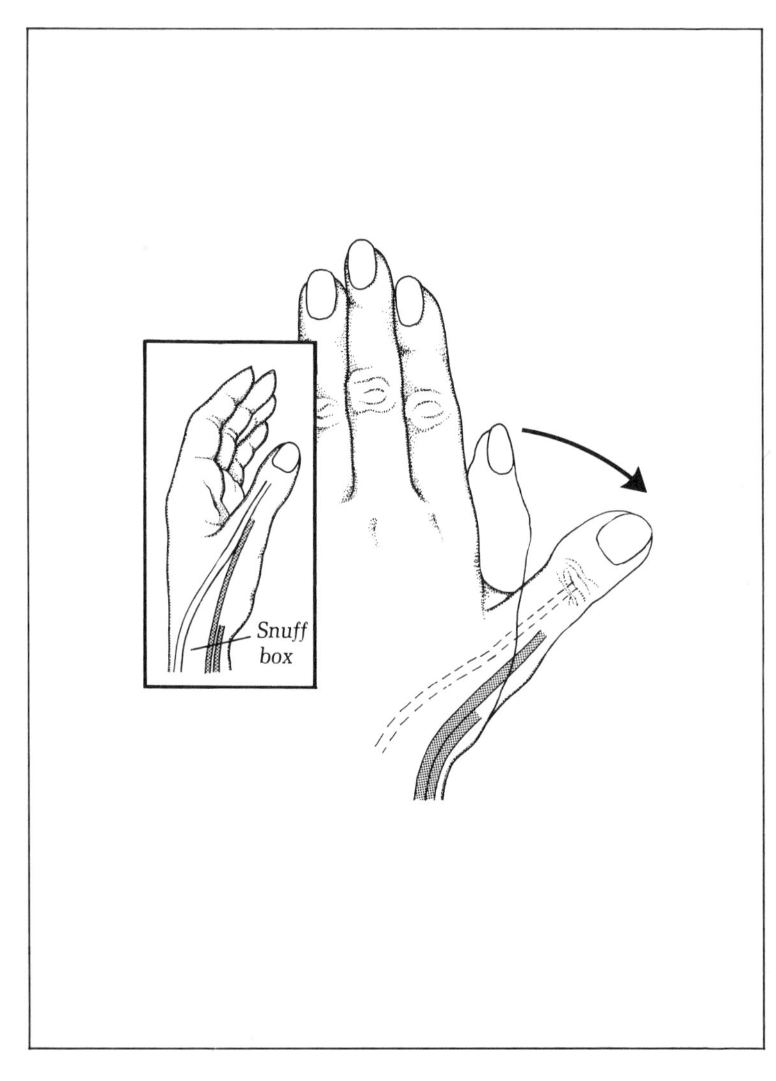

Figure 8
Testing for EPB and APL musculotendinous function

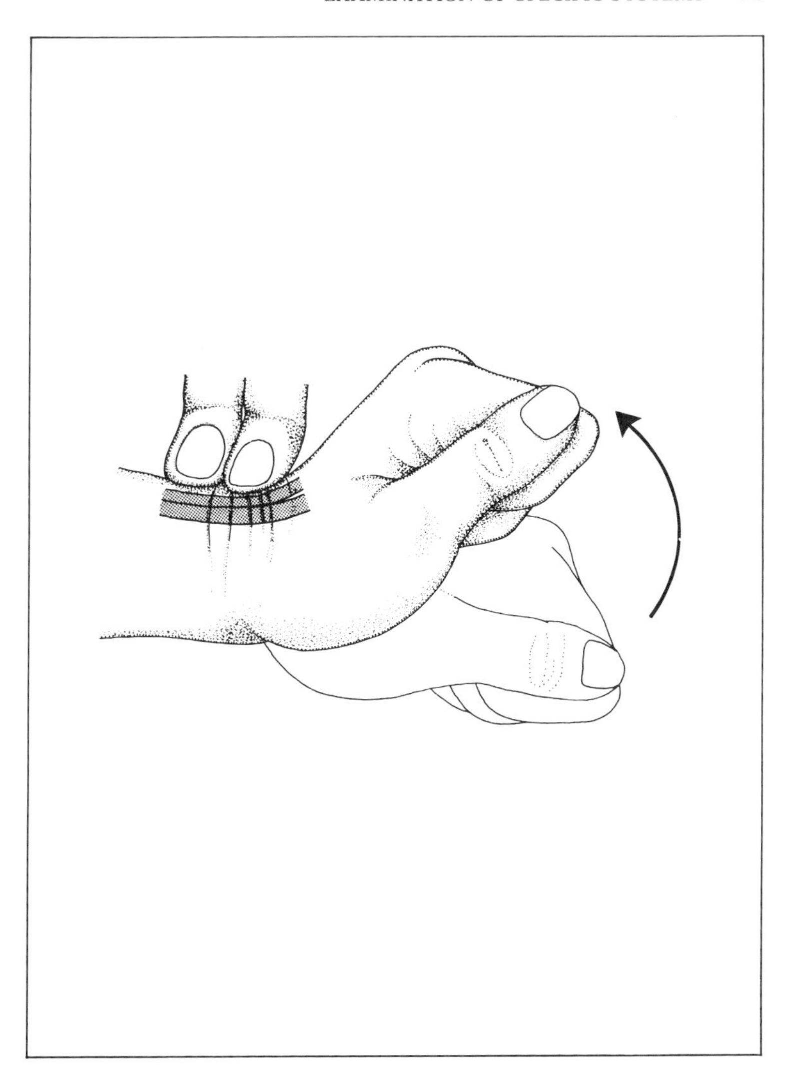

Figure 9
Testing for ECRL and ECRB musculotendinous function

respectively. These are evaluated by asking the patient to 'make a fist and bring your wrist back strongly'. The examiner can give resistance and palpate the tendons over the dorsoradial aspect of the wrist.

The third dorsal wrist compartment. In the third dorsal wrist compartment, the *extensor pollicis longus* (EPL) tendon passes around Lister's tubercle of the radius and inserts on the dorsal base of the distal phalanx of the thumb. This muscle is evaluated by placing the hand flat on the table and having the patient lift only the thumb off the surface (Fig. 10).

The fourth dorsal wrist compartment. The fourth dorsal wrist compartment contains the tendons which are the MCP joint extensors of the fingers (Figs. 7 and 11). The *extensor digitorum communis* (EDC) and the *extensor indicis proprius* (EIP) muscle tendons are evaluated by asking the patient to 'straighten your fingers' and by observing MCP joint extension.

The EIP tendon can be isolated on examination by asking the patien to 'bring your pointing finger out straight, with the other fingers bent in a fist'. The EIP is acting alone to extend the index finger MCP joint (Fig. 11).

The fifth dorsal wrist compartment. The fifth dorsal wrist compartment contains the tendon of the *extensor digiti minimi* (EDM) (Fig. 11). This is evaluated by asking the patient to 'straighten out your little finger with your other fingers bent in a fist'. This extends the MCP joint of the little finger. The EDM is acting alone to extend the little finger.

The sixth dorsal wrist compartment. The sixth dorsal compartment contains the tendon of the *extensor carpi*

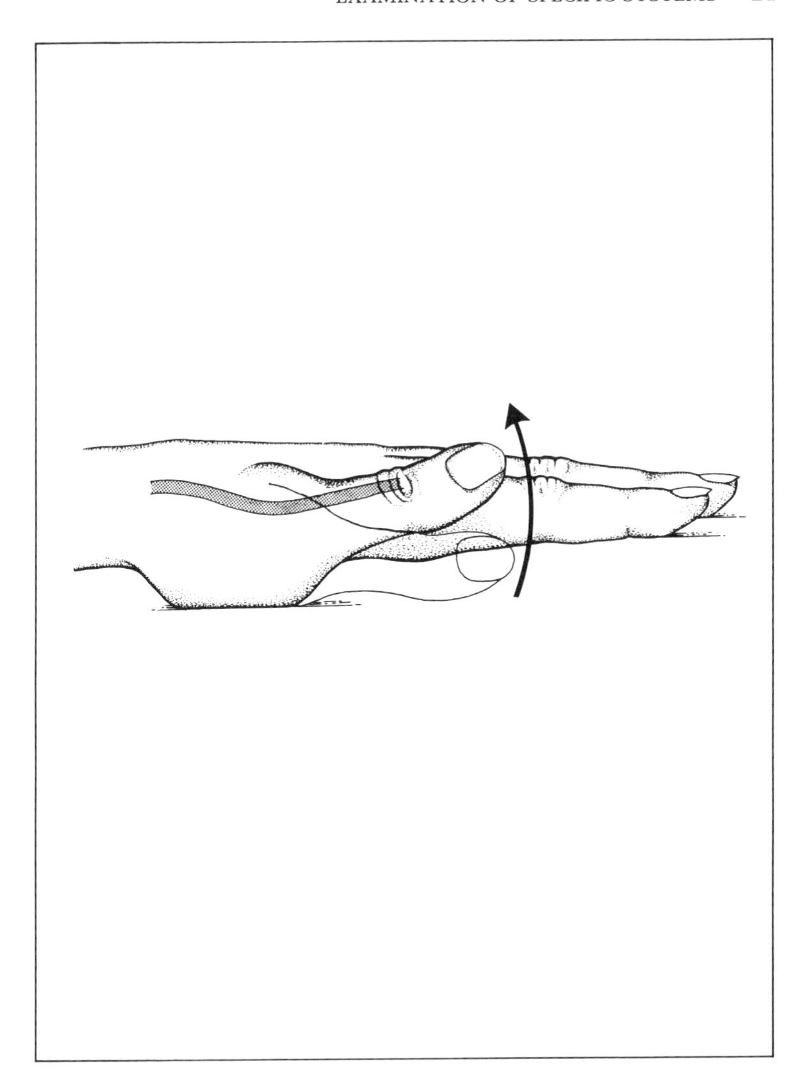

Figure 10
Testing for EPL musculotendinous function

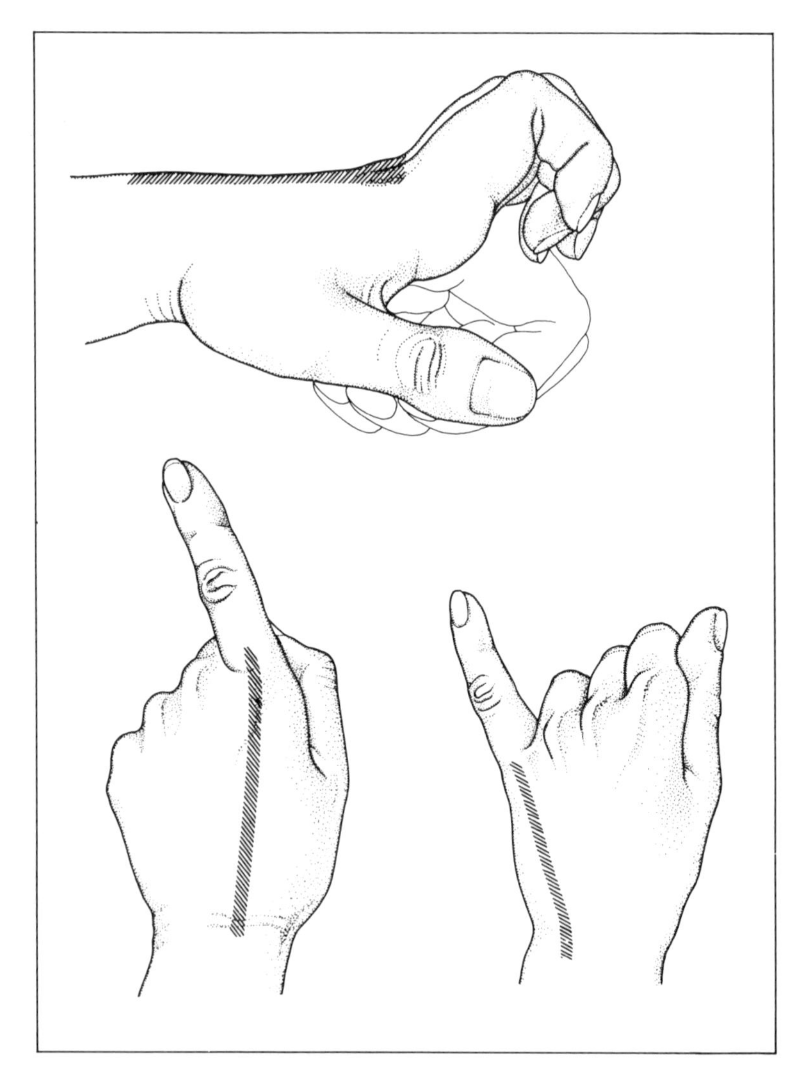

Figure 11
Testing for EDC, EIP, and EDM musculotendinous function

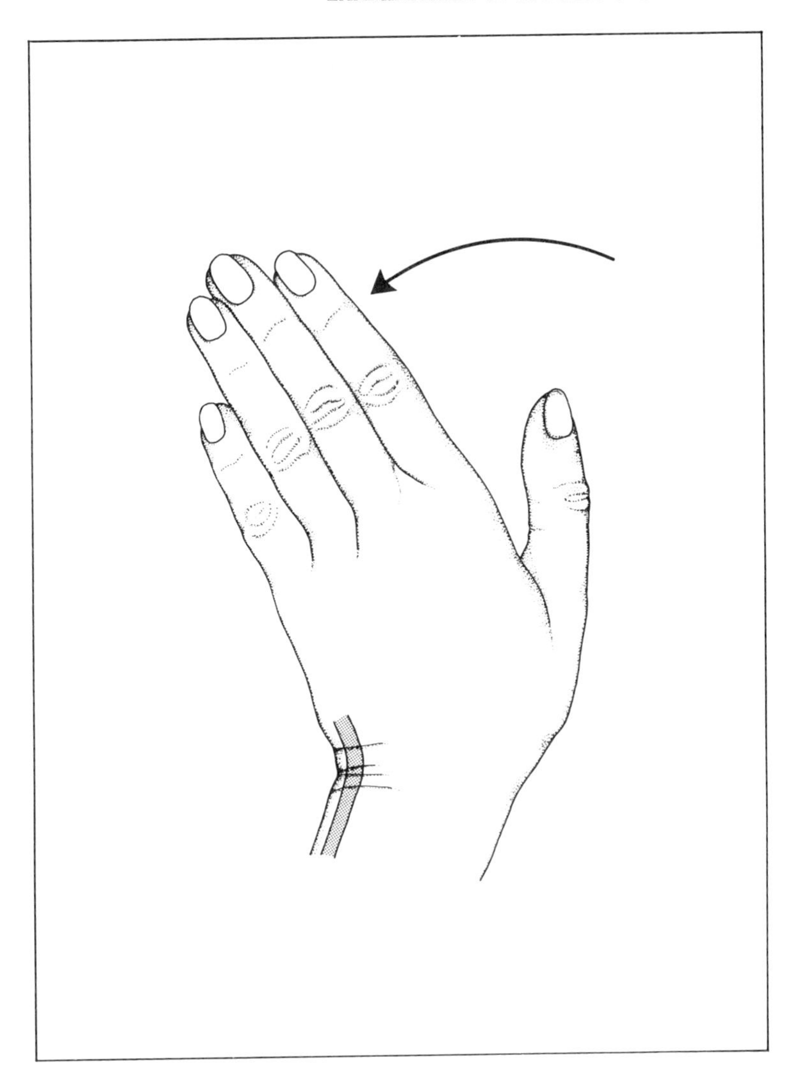

Figure 12
Testing for ECU musculotendinous function

ulnaris (ECU), which inserts at the dorsal base of the fifth metacarpal (Fig. 12). This is evaluated by asking the patient to 'pull your hand up and out to the side'. The taut tendon can be palpated over the ulnar side of the wrist just distal to the ulnar head.

Extrinsic extensor tightness. The extensor tendons can become adherent over the dorsum of the hand or wrist, limiting finger flexion. This can be tested by maintaining the wrist in neutral and passively extending the MCP joint and flexing the PIP joint. Normally, the PIP joint should flex. The test is then repeated with the MCP joint passively flexed. If the PIP joint will passively flex when the MCP joint is extended, but will not flex readily with the MCP joint flexed, the adherent extrinsic extensors are checkreining the simultaneous flexion of finger MCP and PIP joints. This is called 'extrinsic extensor tightness'.

Intrinsic muscles

The intrinsic muscles of the hand are those that have their origins and insertions within the hand. These are the thenar muscle group, adductor pollicis (AdP), lumbrical, and interosseous muscles and the hypothenar muscle group.

The thenar muscles

The thenar muscles are the muscles covering the thumb metacarpal. They are the *abductor pollicis brevis* (APB), *opponens pollicis* (OP), and *flexor pollicis brevis* (FPB). These muscles pronate or oppose the thumb (see Figure 3) and can be evaluated by asking the patient to 'touch the thumb and little finger tips together so that the nails are parallel' (Fig. 13). They can also be tested by asking the patient to place the dorsum of the hand flat on the table and raise the thumb up straight to form a 90 degree angle with the palm (Fig. 3B). At that time it is most important to palpate the thenar muscles to note if they contract. It is helpful to examine and compare the contralateral hand in a similar way to detect slight variations in muscle mass and function. The thenar muscles are usually innervated by the motor branch of the median nerve. In some patients, however, the thenar muscles may be partially innervated by the ulnar nerve.

The adductor pollicis muscle

Thumb adduction is separately tested by having the patient forcibly hold a piece of paper between the thumb and radial side of the index proximal phalanx (Fig. 14). The muscle that powers this motion is the AdP, which is innervated by the ulnar nerve. When this muscle is weak or nonfunctioning, the thumb IP joint flexes with this maneuver (Froment's sign). In this evaluation the two hands must be compared.

The interosseous and lumbrical muscles

The interosseous and lumbrical muscles act on the fingers to flex the MCP joints and extend the IP joints. The interosseous muscles also abduct and adduct the fingers. The interosseous muscles, which lie on either side of the finger metacarpals, are innervated by the ulnar nerve. They can be evaluated by asking the patient to 'spread your fingers apart' while the examiner palpates the first dorsal interosseous to see if it contracts. In another test, with the hand flat on a

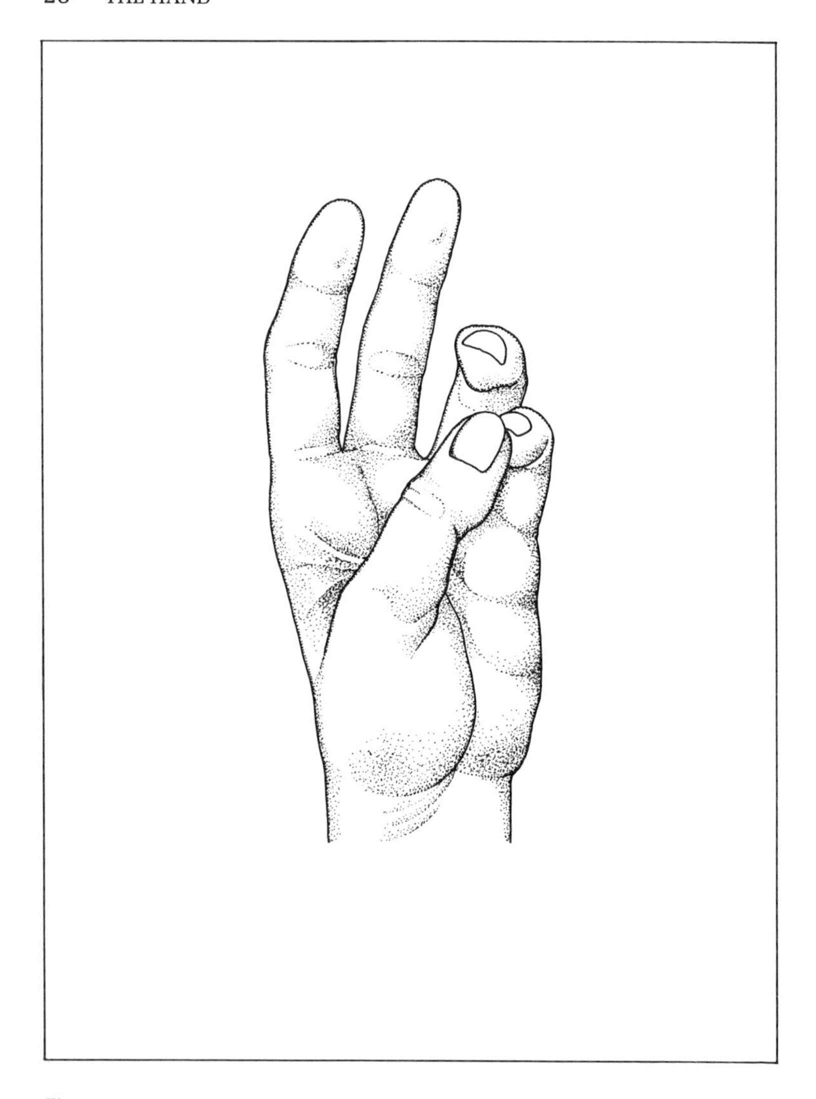

Figure 13
Testing for thumb opposition

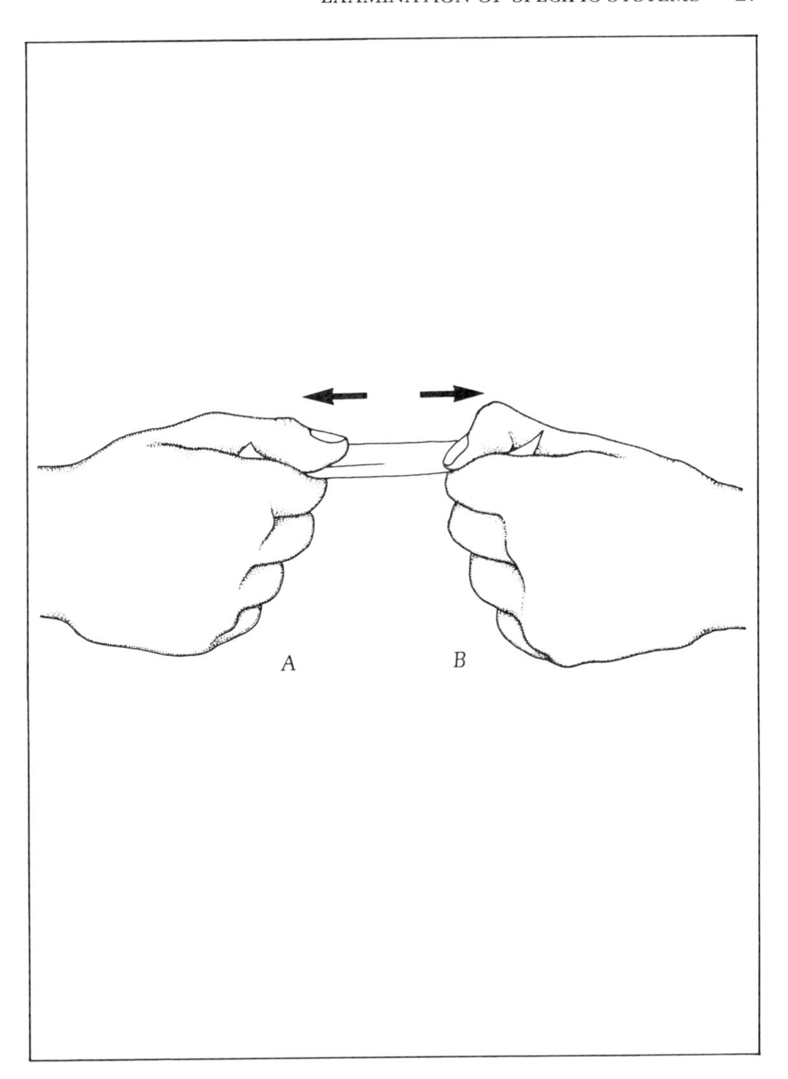

Figure 14
Froment's sign is positive in hand B

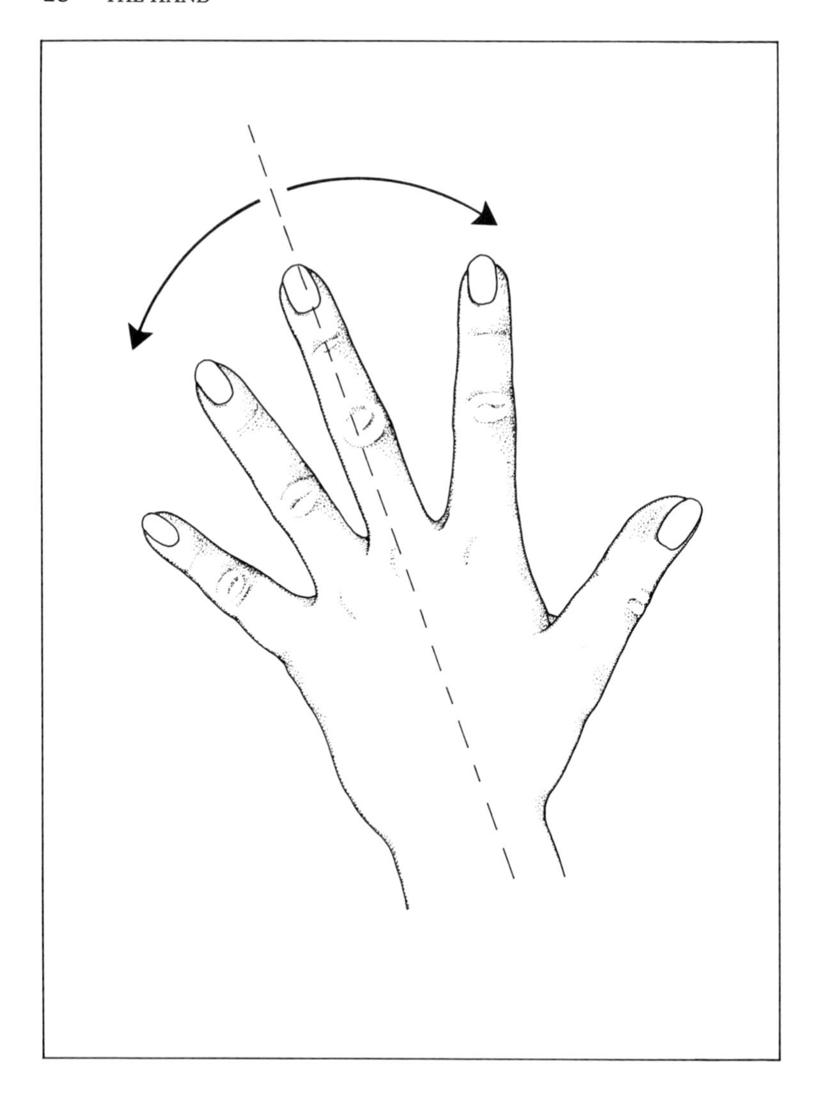

Figure 15
Testing for interosseous muscle function

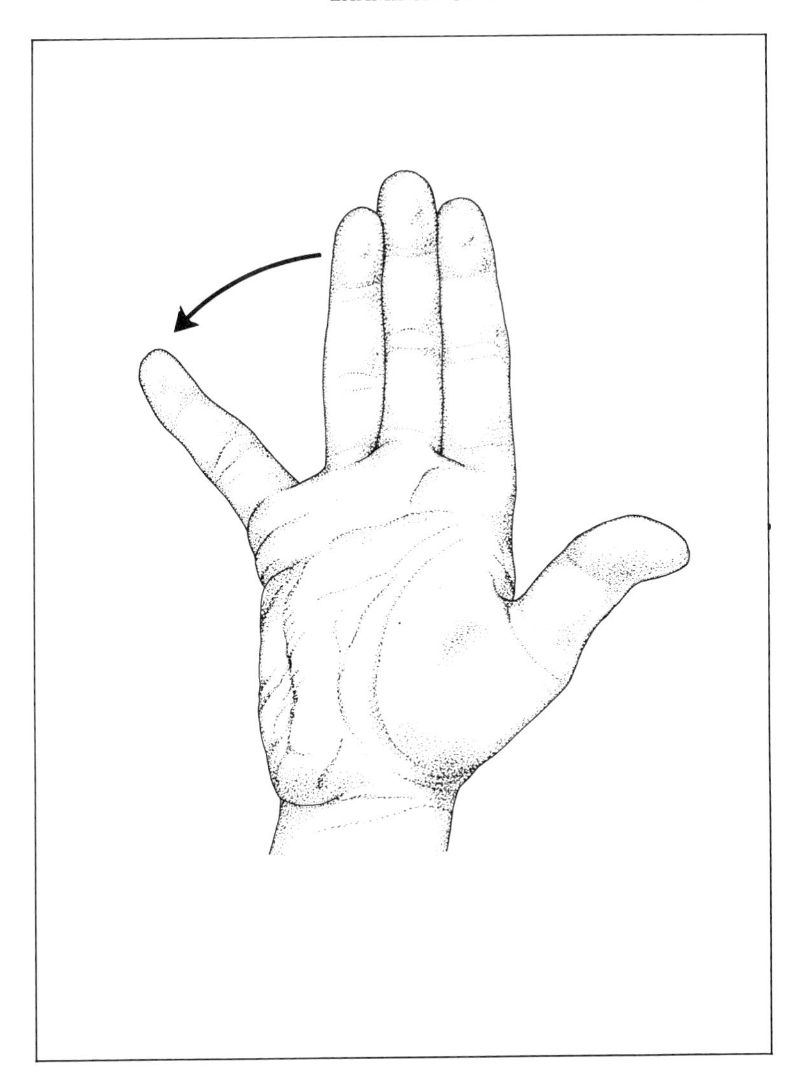

Figure 16
Testing for hypothenar muscle function

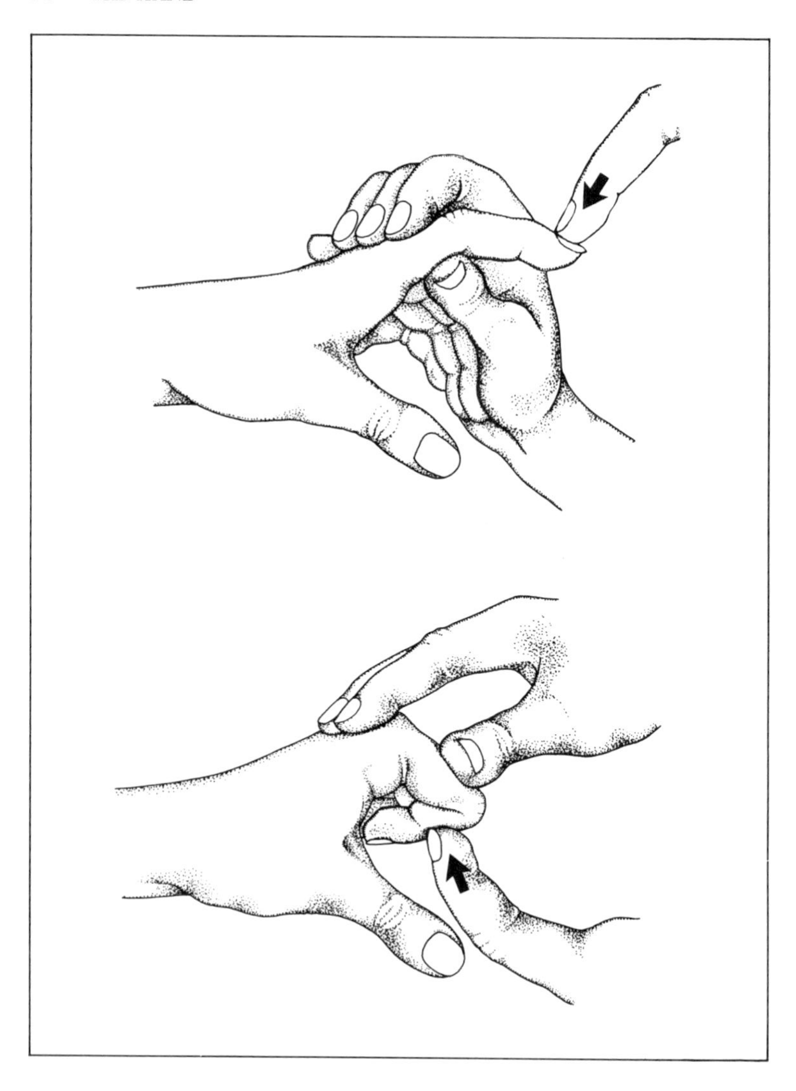

Figure 17
Intrinsic muscle tightness

table, the patient is asked to elevate (i.e., hyperextend the MCP joint with the IP joints straight) the middle finger and radially and ulnarly deviate it (Fig. 15). (This eliminates the extrinsic extensors, which some patients can use to mimic interossei finger abduction-adduction).

The hypothenar muscles

The hypothenar muscles — *abductor digiti minimi* (ADM), *flexor digiti minimi* (FDM), and *opponens digiti minimi* (ODM) — are evaluated as a group by asking the patient to 'bring the little finger away from the other fingers' (Fig. 16). This muscle mass is palpated at that time, and a dimpling of the hypothenar skin noted.

Intrinsic muscle tightness

To test for finger intrinsic muscle tightness the MCP joint of the finger is held in extension (0 degree neutral position) while the PIP joint is passively flexed by the examiner (Fig. 17). The MCP joint is then flexed and the PIP joint passively flexed in the same manner as before. If the PIP joint can be passively flexed with the MCP joint in flexion but cannot be fully flexed when the MCP joint is extended, then there is tightness of the intrinsic muscles. This is called 'intrinsic tightness'.

NERVES

The hand is innervated by the median, ulnar and radial nerves. Each of the three major nerves passes through a muscle in the forearm and each passes points of potential entrapment. All three nerves are involved in control of the wrist, fingers, and thumb.

The median nerve

The median nerve enters the forearm through the pronator teres muscle and innervates the following muscles: pronator teres, flexor carpi radialis, palmaris longus, flexor digitorum superficialis, radial part of the flexor digitorum profundus, flexor pollicis longus, and pronator quadratus (Fig. 18). It enters the hand through the carpal tunnel accompanied by the 9 extrinsic flexor tendons of the digits. The thenar motor branch innervates the abductor pollicis brevis, the lateral half of the flexor pollicis brevis (variably so), and the opponens pollicis. The common digital branches innervate the lumbrical muscles to the index and long fingers. The nerve then continues through the palm as sensory branches (described below).

The ulnar nerve

The ulnar nerve enters the forearm from posterior to the medial epicondyle of the humerus and passes between the two heads of the flexor carpi ulnaris muscle (Fig. 19). It innervates the following muscles in the forearm: the flexor carpi ulnaris and the ulnar part of the flexor digitorum profundus (usually to the ring and little fingers, occasionally to the long finger). It enters the hand at the wrist accompanied by the ulnar artery through a tunnel radial to the pisiform bone, ulnar to the hook of the hamate, volar to the deep transverse carpal ligament, and dorsal to the volar carpal ligament (Fig. 20). This tunnel is known as the ulnar tunnel or Guyon's canal. The ulnar nerve innervates the hypothenar muscles (the abductor digiti minimi, flexor digiti minimi, opponens digiti minimi), the seven interosseous muscles, the lumbrical muscles to the ring and little fingers, and the adductor pollicis. It may innervate part or all of the flexor pollicis brevis.

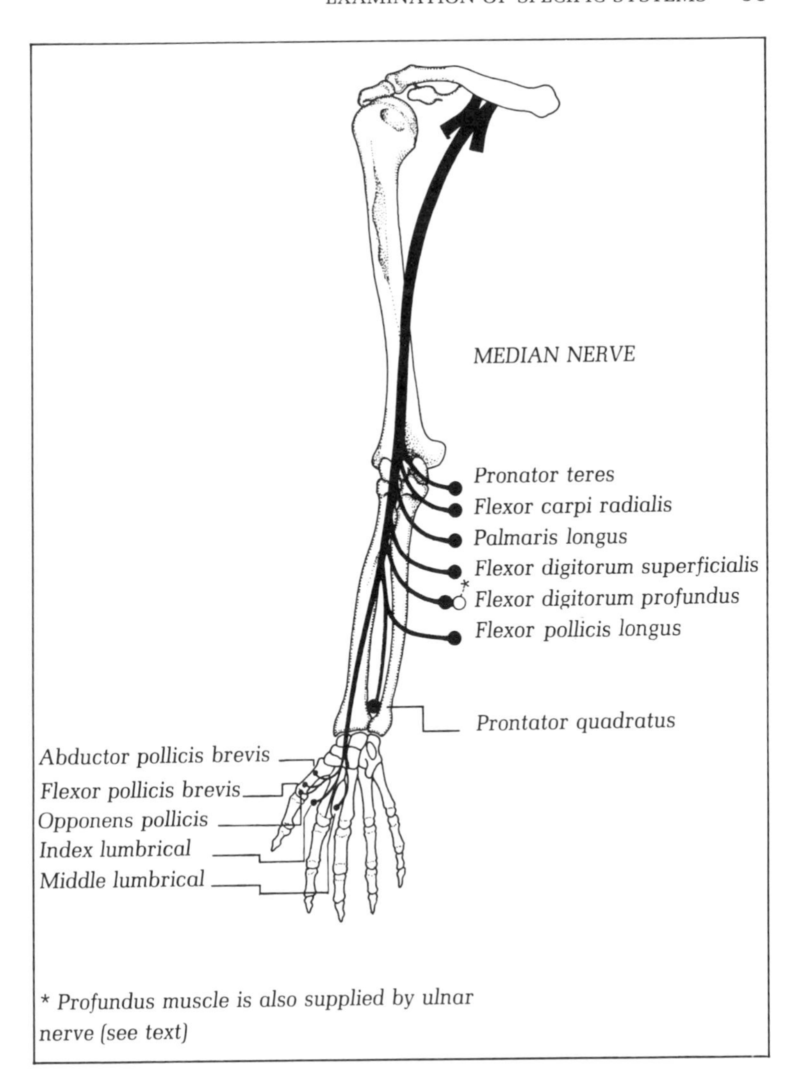

Figure 18
Muscles innervated by the median nerve in the forearm and hand

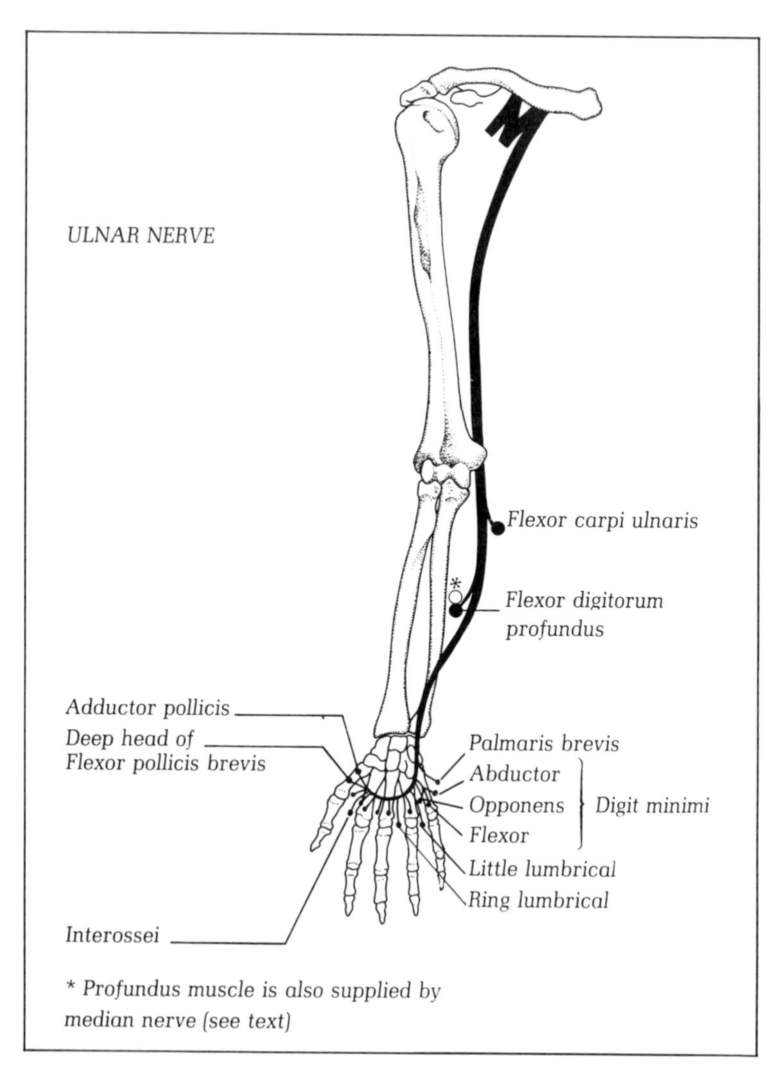

Figure 19
Muscles innervated by the ulnar nerve in the forearm and hand

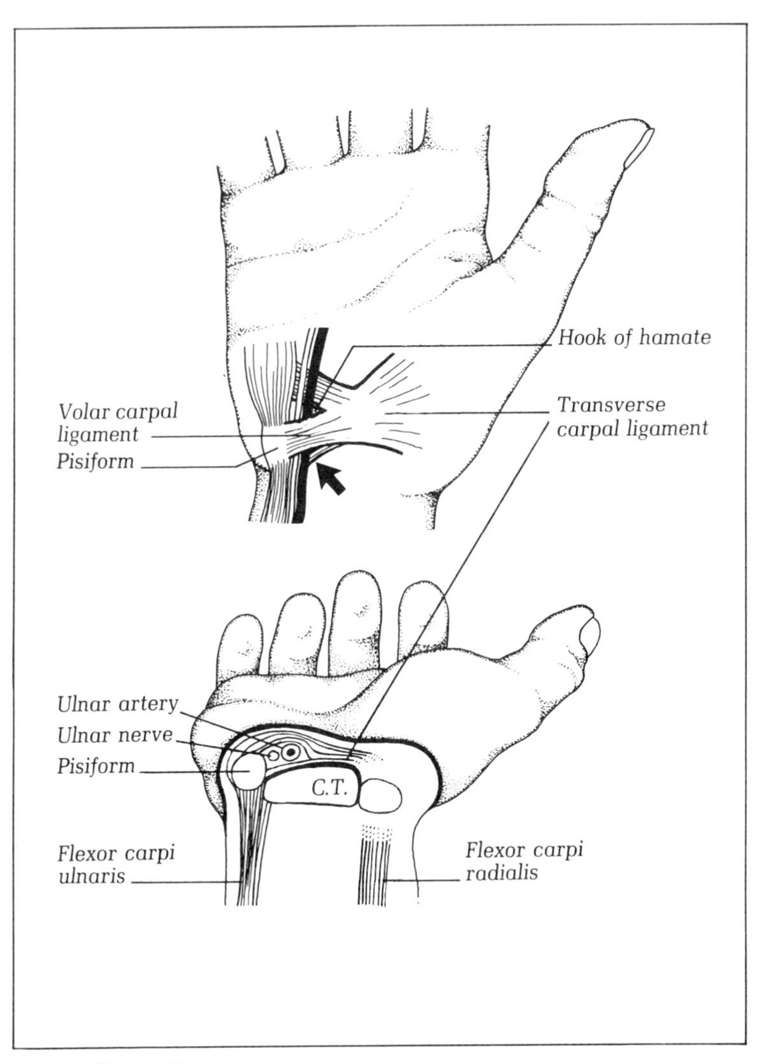

Figure 20
Ulnar tunnel at wrist (Guyon's canal) contains ulnar artery and nerve
Carpal tunnel (CT)

The radial nerve

The radial nerve innervates the triceps, anconeus, brachioradialis, and extensor carpi radialis longus muscles above the elbow and extensor carpi radialis brevis as the nerve enters the forearm (Fig. 21). It passes through the supinator muscle to innervate the following muscles in the forearm: supinator, extensor digitorum communis, extensor digiti minimi, extensor carpi ulnaris, abductor pollicis longus, extensor pollicis longus, extensor pollicis brevis, and extensor indicis proprius. Thus its important motor function is to innervate the muscles in the forearm which extend the wrist and MCP joints and which abduct and extend the thumb. No intrinsic muscles in the hand are innervated by the radial nerve.

Sensory branches of the nerves

The median nerve branches as it leaves the carpal tunnel into common sensory branches which subsequently divide and innervate the volar surface of the thumb, the index and middle fingers, and the radial side of the ring finger (Fig. 22). Dorsal digital branches arise from the digital branches to innervate distal to the PIP joint the dorsal aspect of the index and middle fingers and the radial half of the ring finger.

The ulnar nerve divides distal to the hook of the hamate into digital branches and innervates the little finger and the ulnar half of the ring finger (Fig. 22). The dorsal cutaneous branch of the ulnar nerve enters the dorsal aspect of the hand just distal to the ulnar styloid and innervates the dorsal aspect of the hand over the little and ring metacarpals, the dorsum of the little finger, and the dorsoulnar half of the ring finger.

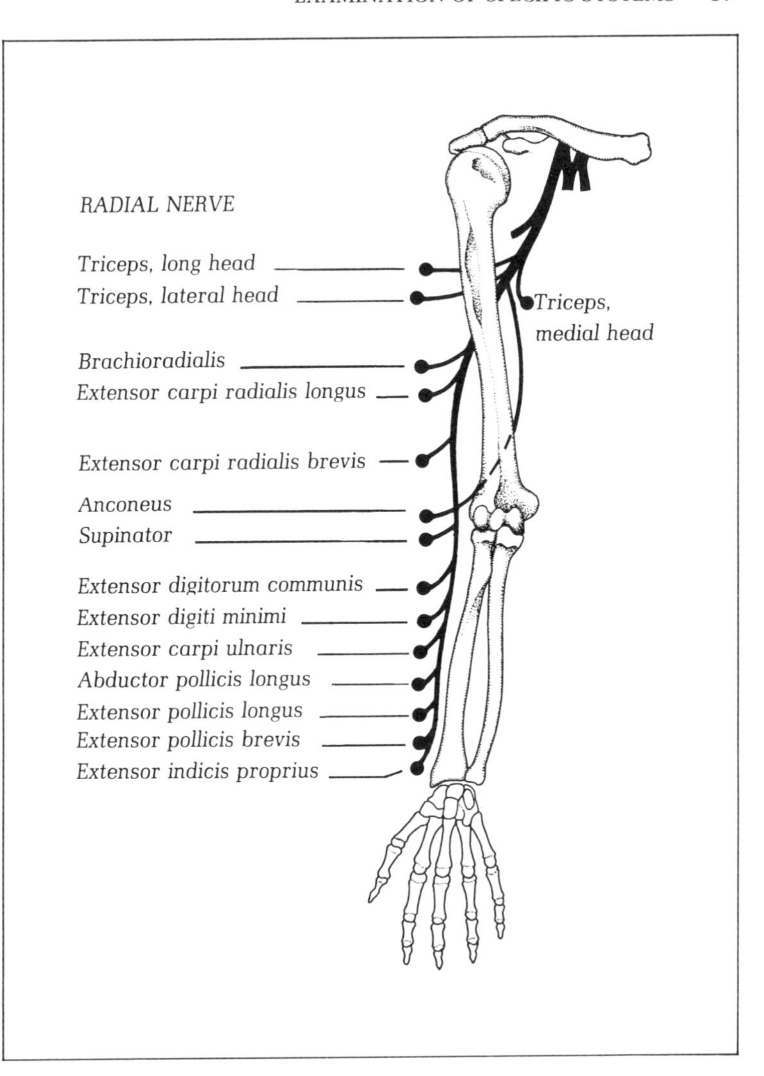

Figure 21
Muscles innervated by the radial nerve in the forearm and hand

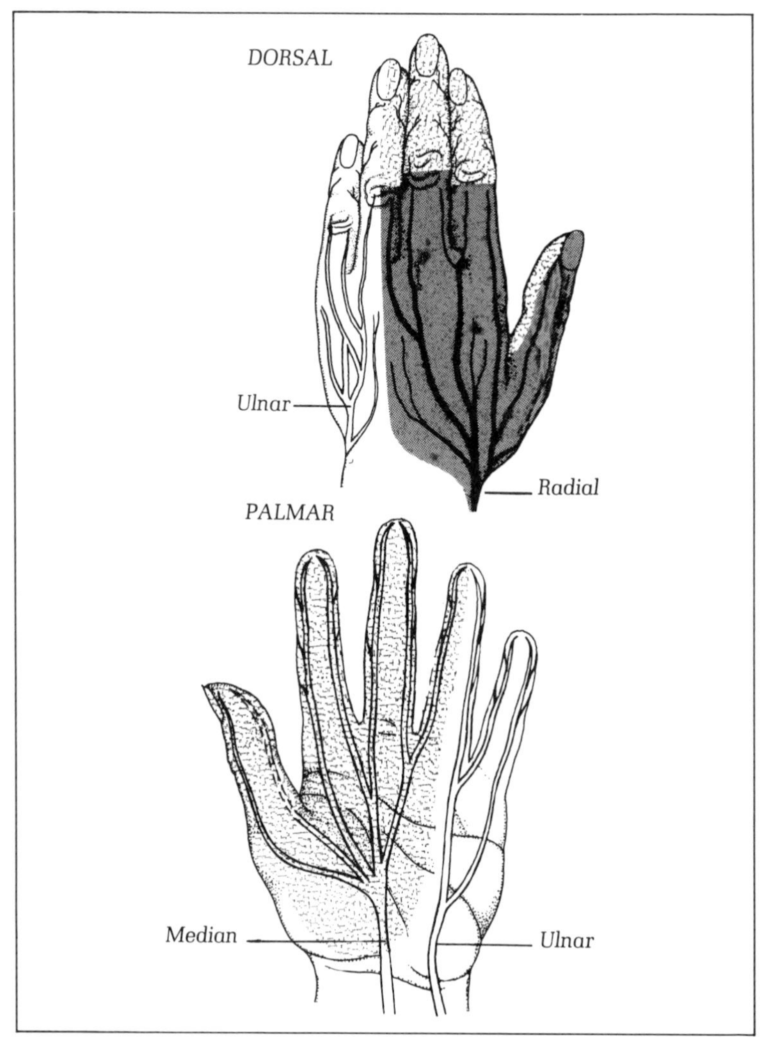

Figure 22
Distribution of major nerves innervating the hand for sensory function

The radial nerve supplies sensibility to the radial three-quarters of the dorsum of the hand and the dorsum of the thumb (Fig. 22). It also supplies sensibility to the dorsum of the index and middle fingers and the radial half of the ring finger as far distally as the PIP joint of each.

Anatomical variation

Anatomical variation should be considered in all cases where there has been an injury to a major nerve trunk. For example, there can be variations in the distribution of the ulnar and median nerves in the hand. The entire ring finger and ulnar side of the long finger may be innervated by the ulnar nerve, or the entire ring finger may be innervated by the median nerve.

Sensibility

Sensibility is one of the most important functions of the hand. The insensible hand is poorly used even when the tendons and joints are normal.

Normal skin should be slightly moist. Nerve dysfunction causes loss of sympathetic innervation in the area of distribution and the skin becomes dry. This is of clinical help in evaluating nerve dysfunction. Testing the finger with a sharp-pointed object, such as a pin, or thermal testing is not as critical and helpful as to test it for tactile gnosis by the two-point light touch discrimination test. In this test the hand is positioned at rest on a flat firm surface and the patient closes his eyes. A paper clip is opened and bent into a caliper (Fig. 23). The two points of the clip are held wide apart and then lightly touched to the finger in the longitudinal axis on the radial or ulnar side of the digit. The patient

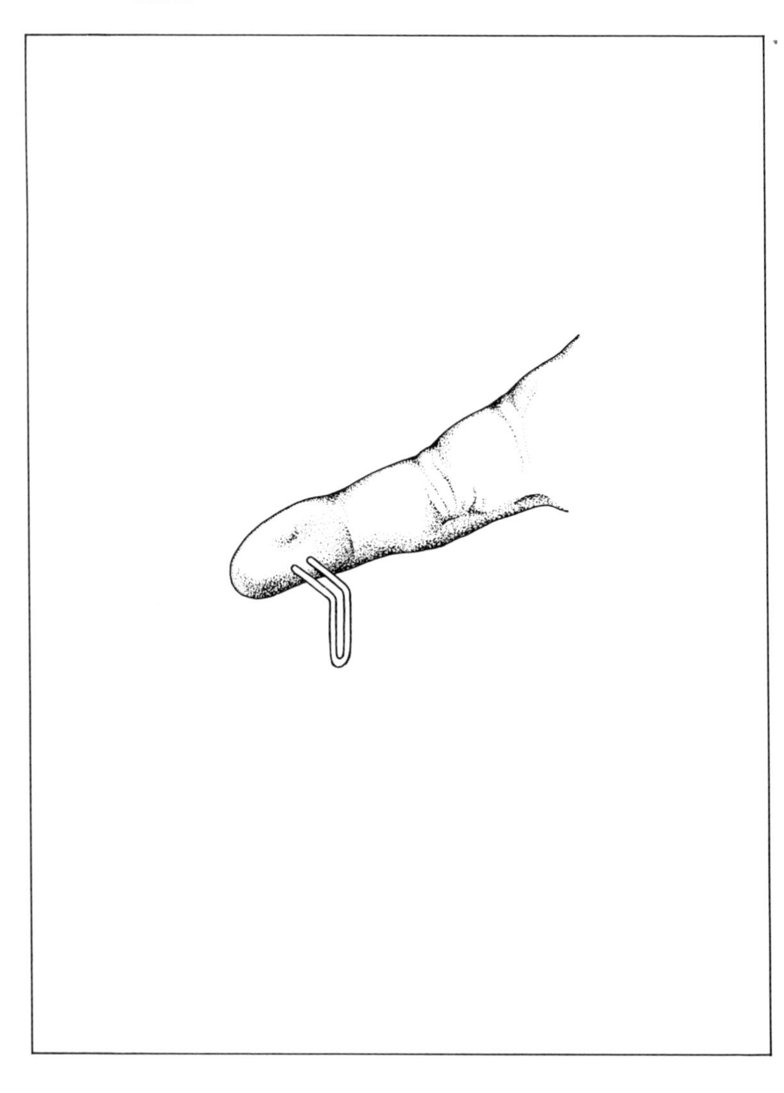

Figure 23
Two-point discrimination testing

indicates whether he feels one or two points. The points are then brought some 2 mm closer together and made to touch the skin again. This action is repeated until the two points feel like one point to the patient; the distance between them is normally less than 6 mm on the tip of the fingers.

CIRCULATION

The radial and ulnar arteries supply the hand with blood. There is an arterial arch system which gives the hand a generous collateral blood supply (Fig. 24).

The circulation of the hand is evaluated by noting the colour of the skin and fingernails as well as the blanching and flush of the nailbed. The Allen test, used to determine patency of the arteries supplying the hand, is done as follows (Fig. 24):

1. Compress the radial and ulnar arteries at the wrist.
2. Have the patient 'make a fist', open and close it several times to exsanguinate the hand, and then open the hand again into a relaxed position (avoid hyperextension at this point, as it will maintain blanching).
3. Release the radial artery only. If the palm and all five digits fill with blood, then the radial artery is patent, with good collateral flow into the ulnar artery system.
4. Repeat steps 1 and 2.
5. Release the ulnar artery only. If the entire hand flushes, then the ulnar artery is patent, with good flow into the radial system.

The Allen test can also be carried out on a single digit by expressing the blood out of the digit and occluding

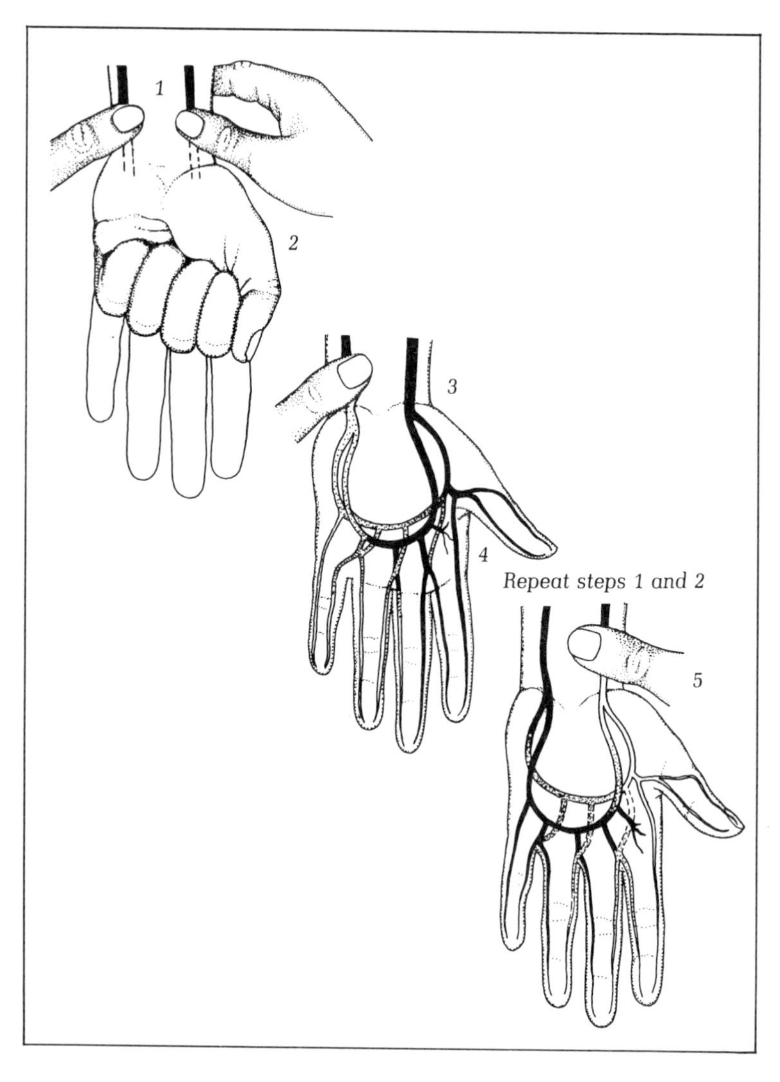

Figure 24
Allen test for arterial patency

both digital arteries and then releasing the radial digital artery and noting the filling of the digit. The same procedure is carried out on the ulnar digital artery. This will help to evaluate the patency of each digital vessel to that finger.

ANATOMY OF THE BONES AND JOINTS

The skeleton of the hand consists of 27 bones, divided into three groups: the carpus, the metacarpal bones, and the phalanges (Fig. 2).

The carpus

The eight carpal bones are divided into two rows. Those in the proximal row, beginning from the radial side, are the scaphoid, lunate, triquetrum, and pisiform. Those in the distal row are the trapezium, trapezoid, capitate, and hamate. Much of the surface of the carpal bones is covered with cartilage, with roughened areas dorsally and volarly for ligamentous attachments and for entry of the vascular supply to the bone.

Wrist flexion and extension as well as radial and ulnar deviation result from radiocarpal and intercarpal motion, whereas pronation and suppination occur through the proximal and distal radioulnar joints.

It is important to emphasize that the hand is not flat. It is based on a system of skeletal arches which must be maintained to preserve hand function (Fig. 25).

Fixed and mobile units

The metacarpals of the index and long fingers are firmly attached to the rigidly interconnected distal

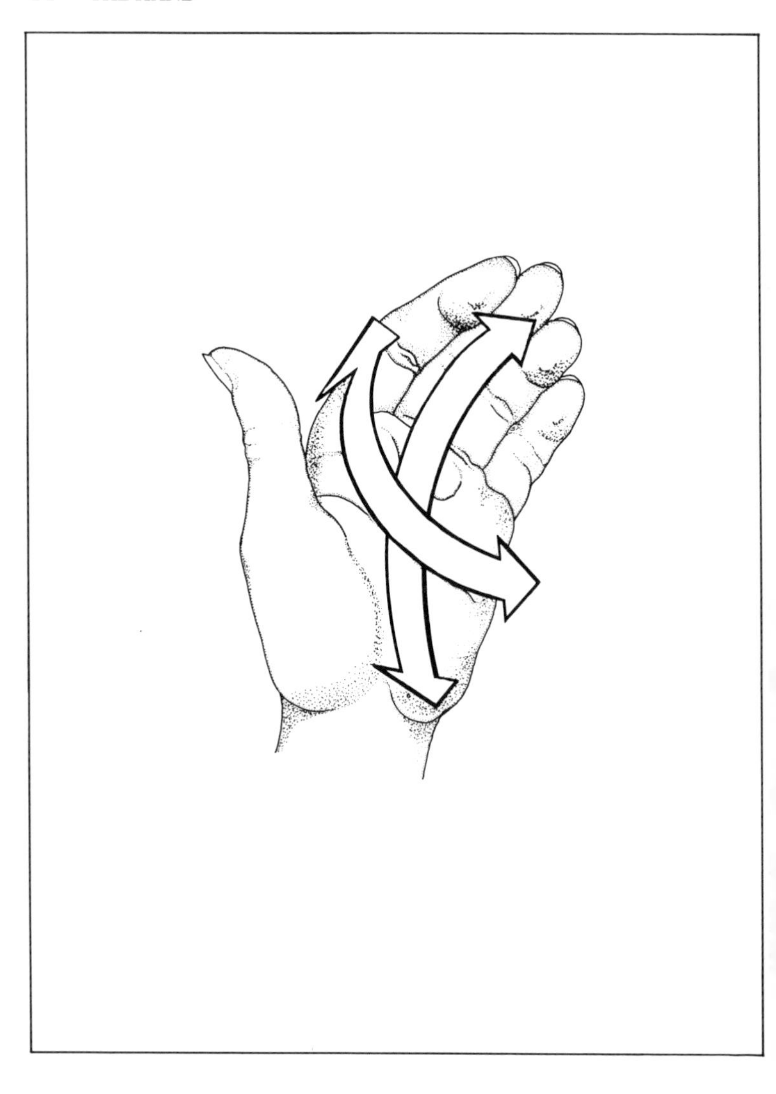

Figure 25
Arches of hand

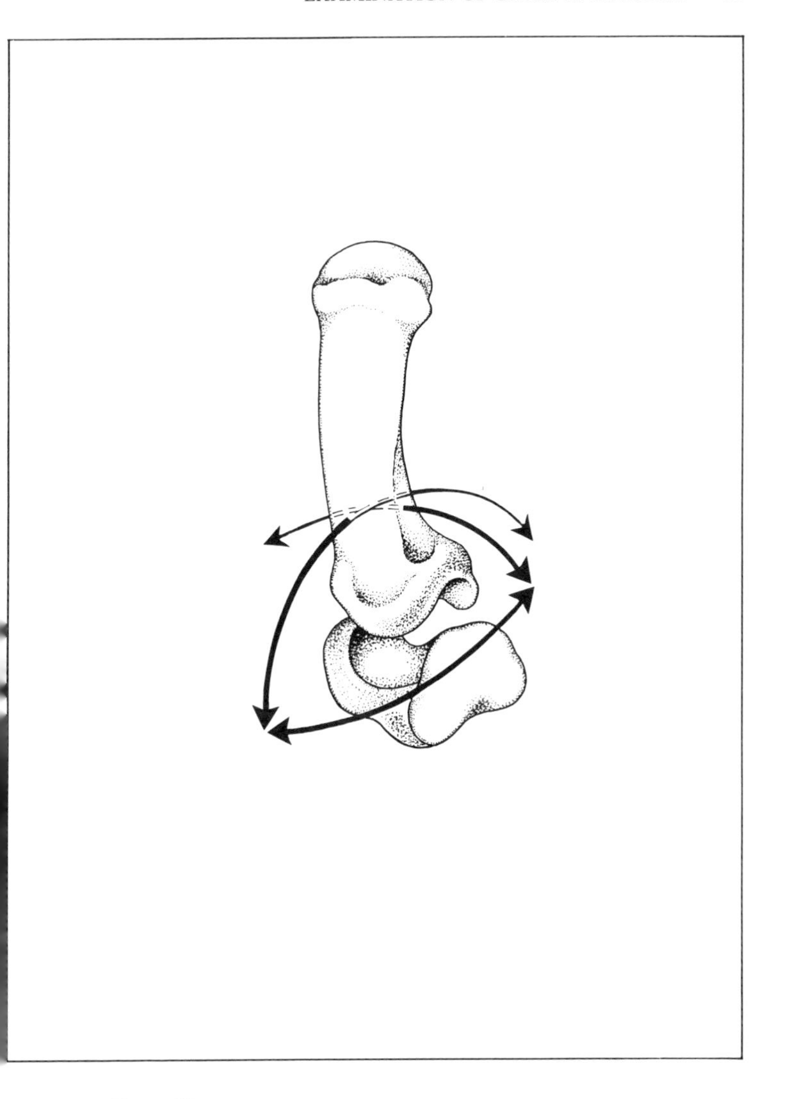

Figure 26
Saddle joint of thumb

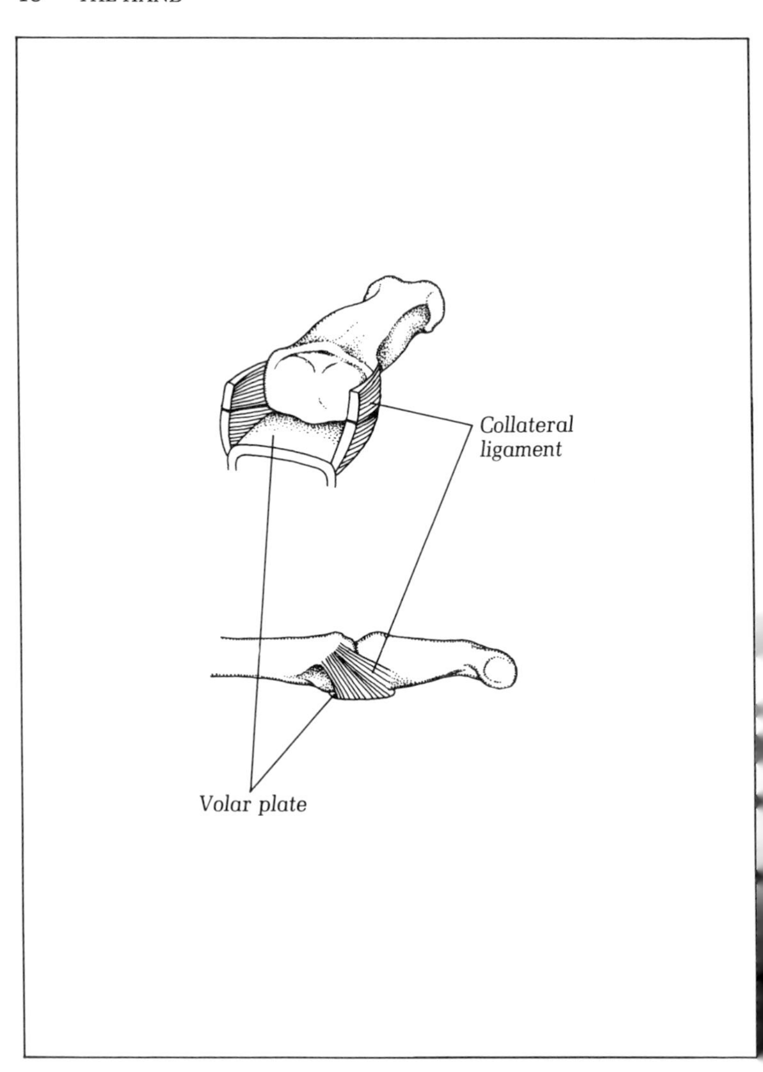

Figure 27
Collateral ligaments and volar plate of the PIP joint

carpal row to form the 'fixed' unit of the hand. From this are suspended the 'mobile or adaptive' components of the hand — the thumb, the entire ring and little rays (including metacarpals), and the phalanges of the index and long fingers.

The longitudinal arch is apparent in the lateral projection and is formed by the metacarpals and phalanges. There are two transverse arches: the proximal arch at the distal carpus and the distal arch at the metacarpal heads.

The thumb metacarpal articulates with the trapezium, forming the unique 'saddle joint', which allows for a wide latitude of thumb motion (Figs. 3B and 26).

The MCP and IP joints of the fingers are stabilized on both sides by collateral ligaments and anteriorly by a palmar fibrocartilaginous 'volar' plates (Fig. 27). The digital flexor tendons lie just anterior to these plates. The configuration of the metacarpal heads causes their collateral ligaments to be slack in extension, permitting abduction, adduction, and circumduction. In flexion, however, MCP collateral ligaments become taut, providing stability to the joint.

Articular configuration of the IP joints and the geometry of the collateral ligaments do not allow significant medial-lateral motion in extension or in flexion. The MCP joint of the thumb is more like the hinged IP joints than the freely movable MCP joints of the fingers.

PART 2

COMMON CLINICAL PROBLEMS

3

LACERATIONS

One should develop a routine for examining the patient with a lacerated forearm or hand so that nerve and tendon injuries will not be overlooked. The patient who presents with a bleeding laceration of the hand should be asked to lie down. The hand is elevated, a sterile dressing is used to cover the wound, and gentle direct pressure is applied. The bleeding will usually stop within a few minutes. The practice of 'clamping a bleeder' in a lacerated hand should be avoided. Previously undamaged vital structures, such as nerve or tendon, may be inadvertently crushed and irrevocably damaged in an unnecessary attempt to clamp a blood vessel.

There is a tendency on the part of inexperienced physicians to look into the wound and see if nerves or tendons have been cut. However, much more can be learned on the initial examination by covering the wound and performing a gentle, systematic examination of the forearm and hand *distal* to the injury (Fig. 28). Each flexor tendon must be tested separately for function, and it is important to test IP function against gentle resistance. A partially cut tendon may be able to flex the finger, but it will not be able to do so against resistance without causing pain.

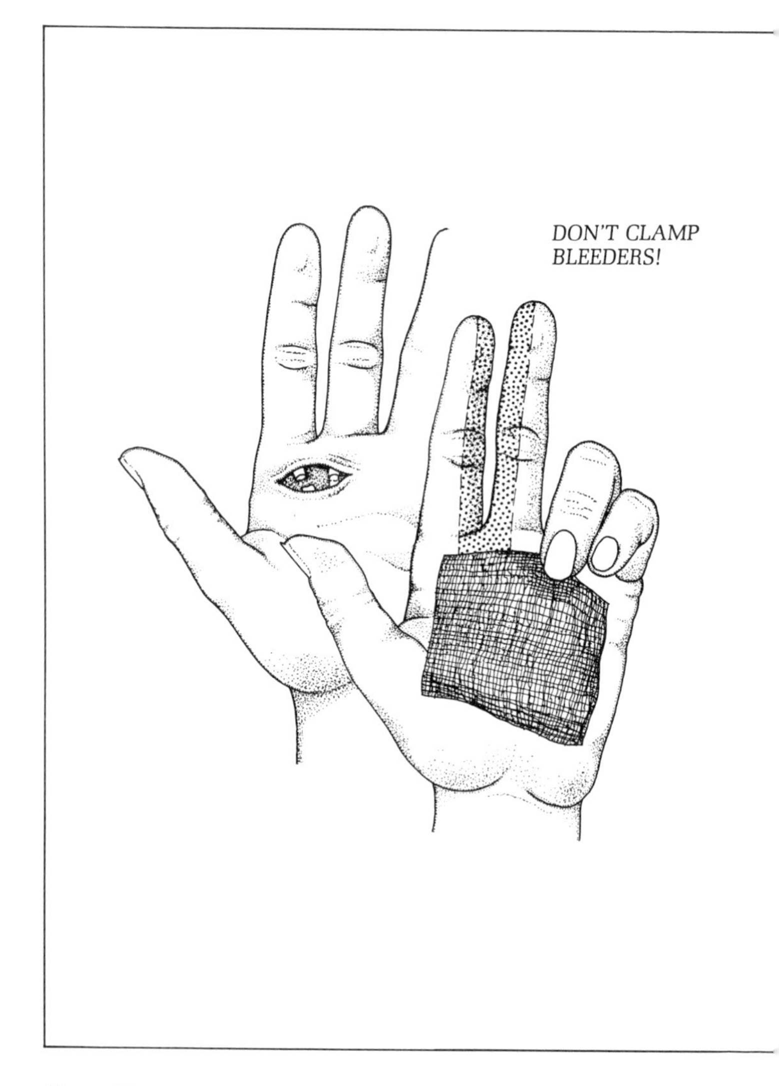

Figure 28
Examination of the lacerated hand

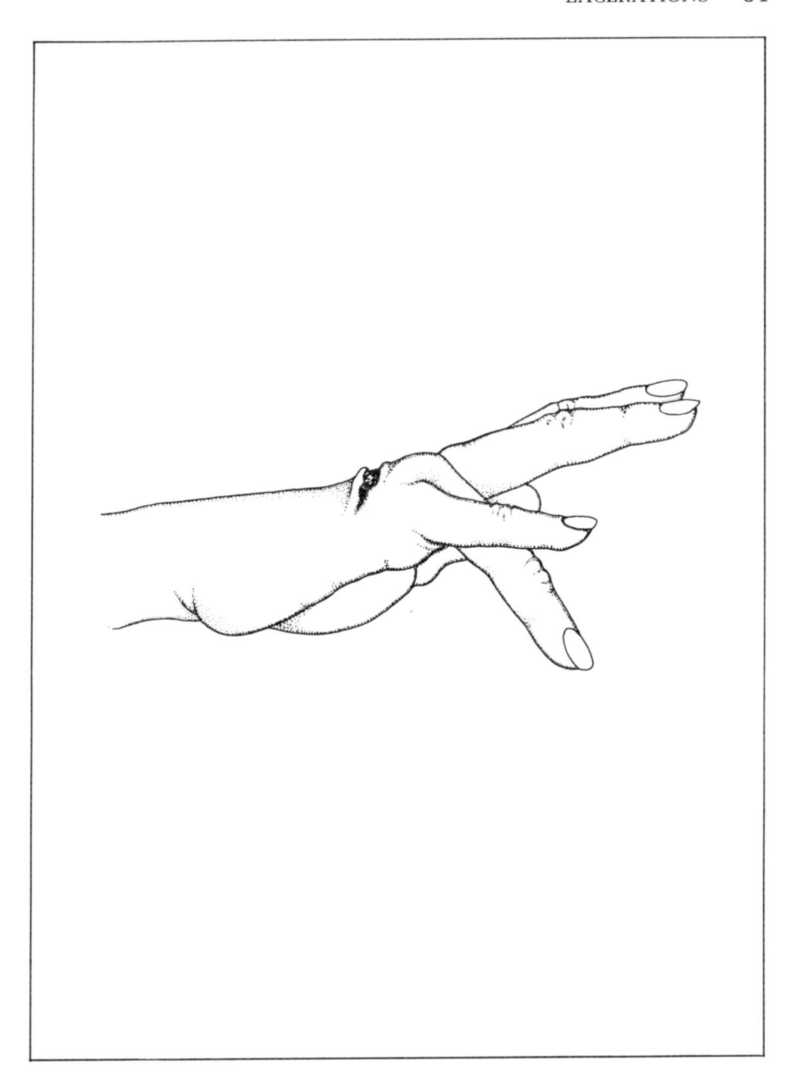

Figure 29
Laceration of EDC over MCP joint (i.e. distal to juncturae tendinae)

The position of the unsupported fingers should be noted. When the flexor tendon is completely severed, the unsupported finger rests in extension (Fig. 28); when the extrinsic extensor tendon is completely severed, the unsupported finger rests in flexion (Fig. 29). A careful distal sensory examination is then done. In the emergency room setting, especially with a frightened child, far more can be learned about the presence or absence of sensation using a light touch with a wisp of cotton than by testing sharp/dull with a pin. Only after the hand has been completely assessed by the examining physician should any anesthetic be used.

In lacerations of the dorsal aspect of the MCP joint of the finger, a severed extensor digitorum communis will preclude active extension of the MCP joint (Fig. 29).

Note that the intact intrinsic muscles will actively extend the IP joints in the absence of extrinsic extensor tendon function, just as the intact intrinsic muscles will actively flex the MCP joint in the absence of extrinsic flexor tendon function.

A laceration over the MCP joint (knuckles) should alert the examiner to the possibility of its having resulted from a human bite or a blow against some teeth. These lacerations are of special importance because of the risk of severe infection.

A roentgenogram of the hand in the anteroposterior, lateral and oblique views should be done. Some glass is radiopaque and will therefore be shown on the film. However, some glass, wood, and plastic may not be radiopaque and may not be seen on the film.

Associated fractures should be ruled out.

4

COMMON FRACTURES AND DISLOCATIONS OF THE HAND

Fractures of the bones of the hand are classified by the nature and site of the fracture line and whether the fracture is closed or open (Fig. 30). An open fracture is one that communicates with the skin wound.

Because of angular or rotational deformity, simple inspection of the hand will often alert the examiner that a bone or joint injury has occurred. It is important that proper antero-posterior and *true lateral* roentgenograms be obtained to confirm the presence of bony injury; the amount of angulation of the fracture may not be appreciated on improperly positioned views. A careful physical examination is essential to evaluate rotational alignment. Since the flexed fingers normally point toward the tubercle of the scaphoid, malrotation is best evaluated by observing the fingers in this position. The rotational alignment of the involved finger can also be compared with that of adjacent uninjured fingers by noting if the planes of the distal fingernails are parallel.

It is important to realize that the deformity of fractures in the hand is due not only to the mechanism of injury but also to the deforming forces of the musculotendinous units acting across the fracture site (Fig. 31).

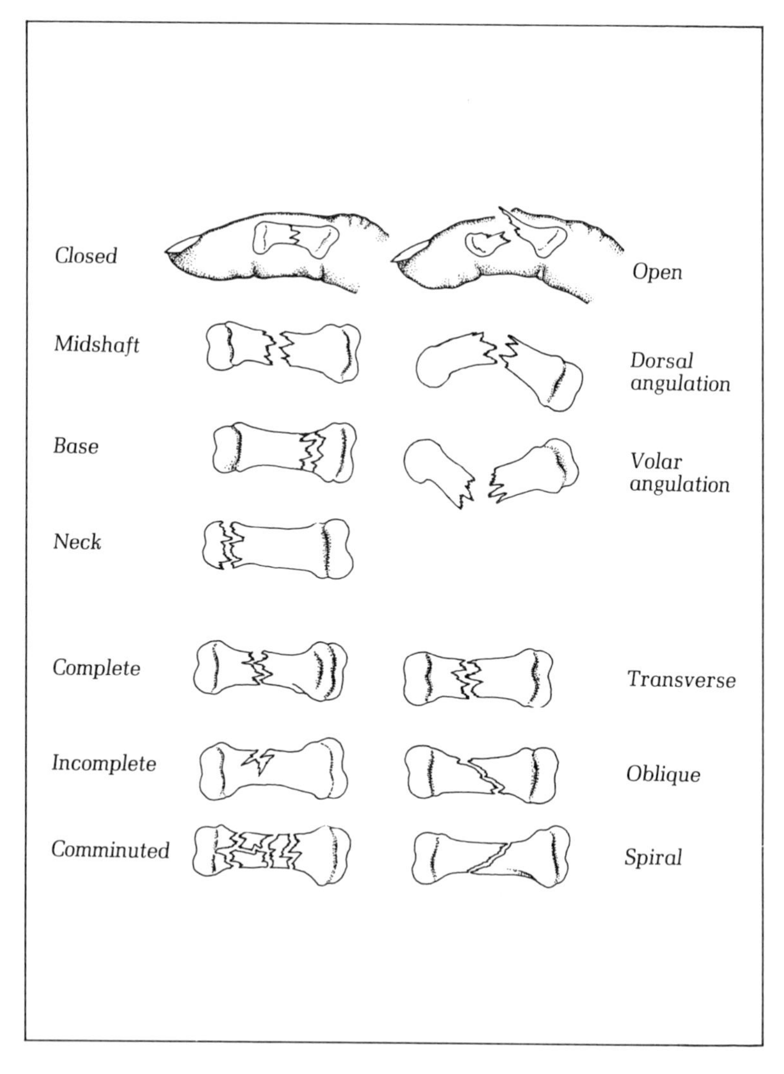

Figure 30
Fracture terminology

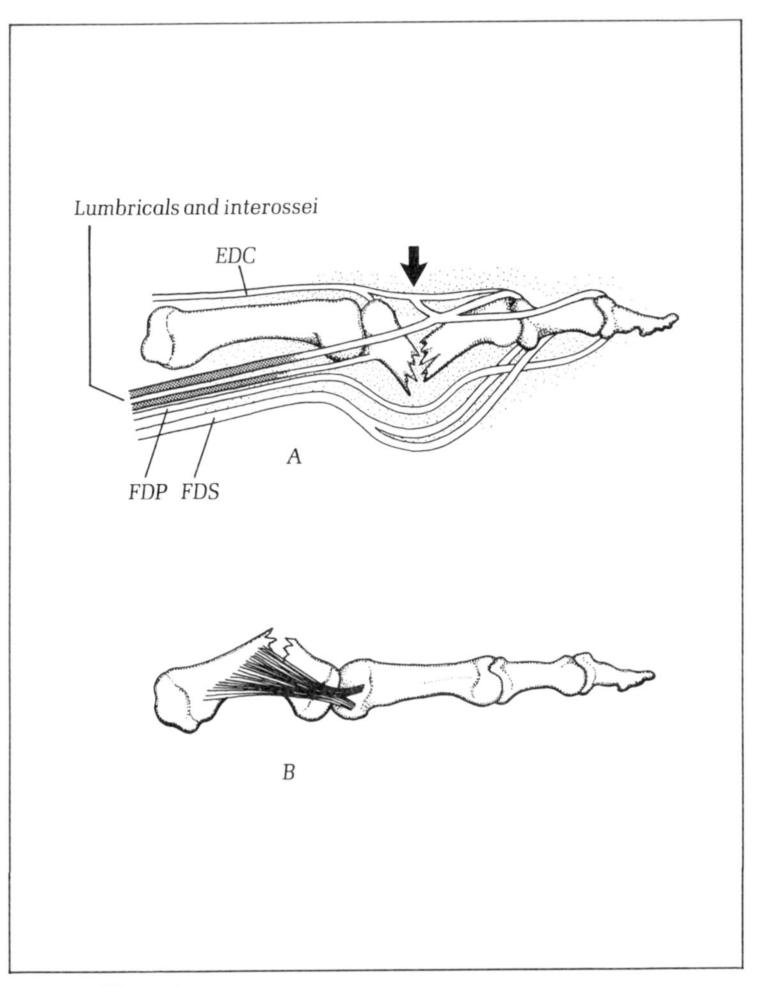

Figure 31
Deforming force acting on fracture site. In (A) the intrinsics flex the proximal fragment of the proximal phalanx. The intrinsic flexor and extensor with longitudinal pull cause further buckling at the fracture site. (B) the intrinsic muscle causes flexion deformity of metacarpal fracture

SPECIFIC FRACTURES AND JOINT INJURIES

It is apparent from the following examples that a knowledge of the functional anatomy of the soft tissues related to the joints of the hand is essential to understanding these potentially disabling injuries. The roentgenograms may misleadingly suggest that a very simple fracture has occurred with only a small fragment of the bone involved. This fragment, however, is often the major attachment of a collateral ligament, the volar plate, or a tendon. This small fracture may render the joint grossly or potentially unstable. Since many of these articular fractures were actually dislocations at the time of injury, the X-ray film may not indicate the true degree of original displacement that occurred.

Intra-articular fractures

Particular attention should be directed to intra-articular fractures around the PIP joint (Fig. 32). These often involve injuries to the volar plate and portions of the collateral ligaments. The early objective evidence of this may be seen on a roentgenogram as small, avulsed fragments of bone around the joint. When a volar triangular fracture fragment from the middle phalanx involves more than one quarter of the articular surface, dorsal dislocation of the middle phalanx *may* occur late because the volar plate and a significant portion of the collateral ligaments are attached to this small fragment. Because of this instability, these fractures often require surgical treatment. Early recognition and proper treatment are dependent upon an awareness of the importance of these initial roentgenographic

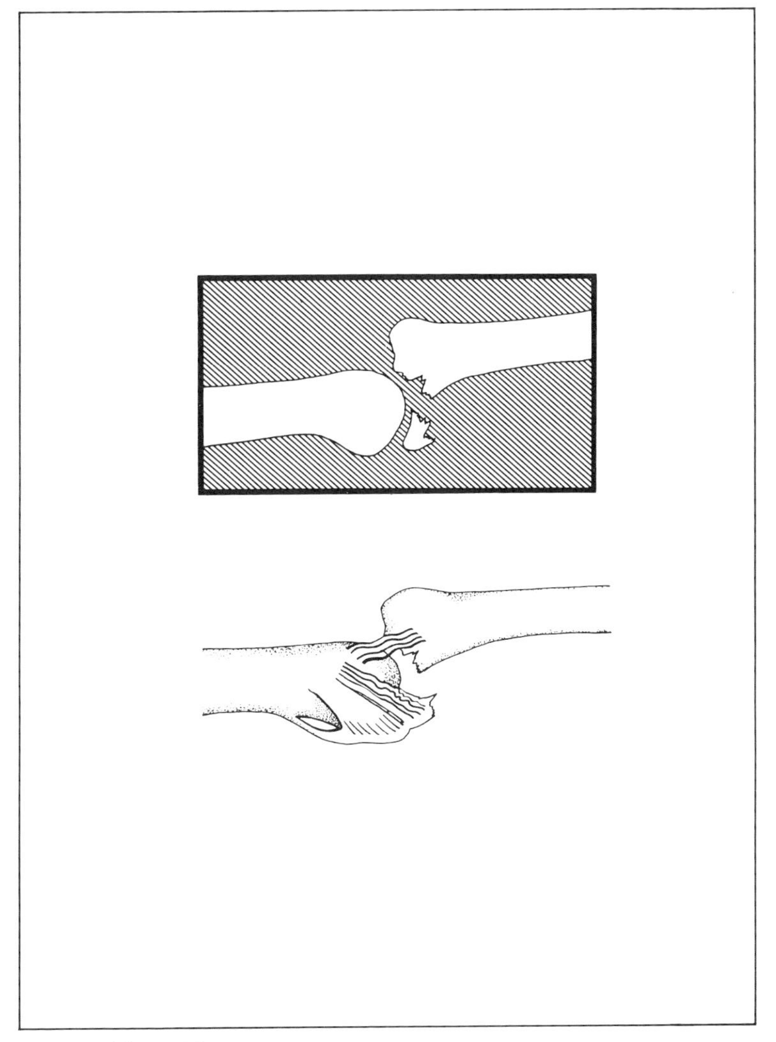

Figure 32
Unstable fracture dislocation of PIP joint with volar fragment, showing X-ray appearance above and ligament attachments below

findings. Undertreatment is a common cause of disability.

Bennett's fracture

Bennett's fracture is an oblique intra-articular fracture from the ulnar base of the thumb metacarpal (Fig. 33). The volar-ulnar portion of the metacarpal, which with its heavy ligamentous attachments normally stabilizes this joint, is separated from the larger distal fragment which is displaced by the pull of the abductor pollicis longus.

'Boxer's fracture'

'Boxer's fracture' usually involves the acute angulation of the head of the metacarpal of the little finger into the palm as the result of a blow on the distal-dorsal aspect of the closed fist. A loss of prominence of the metacarpal head is often seen on physical examination. The active motion of the little finger may be minimally disturbed on initial examination.

Fracture of scaphoid

The bone most commonly fractured in the wrist is the scaphoid. There is tenderness on deep palpation in the snuff-box area of the wrist just distal to the radial styloid (Fig. 34). An oblique roentgenogram ('scaphoid view') will usually best show the fracture. Frequently the initial roentgenogram will fail to show the fracture, whereas repeat views of the scaphoid taken two weeks later may show it after there has been resorption of bone at the fracture site. In all wrist injuries with snuff-box tenderness, careful evaluation with adequate

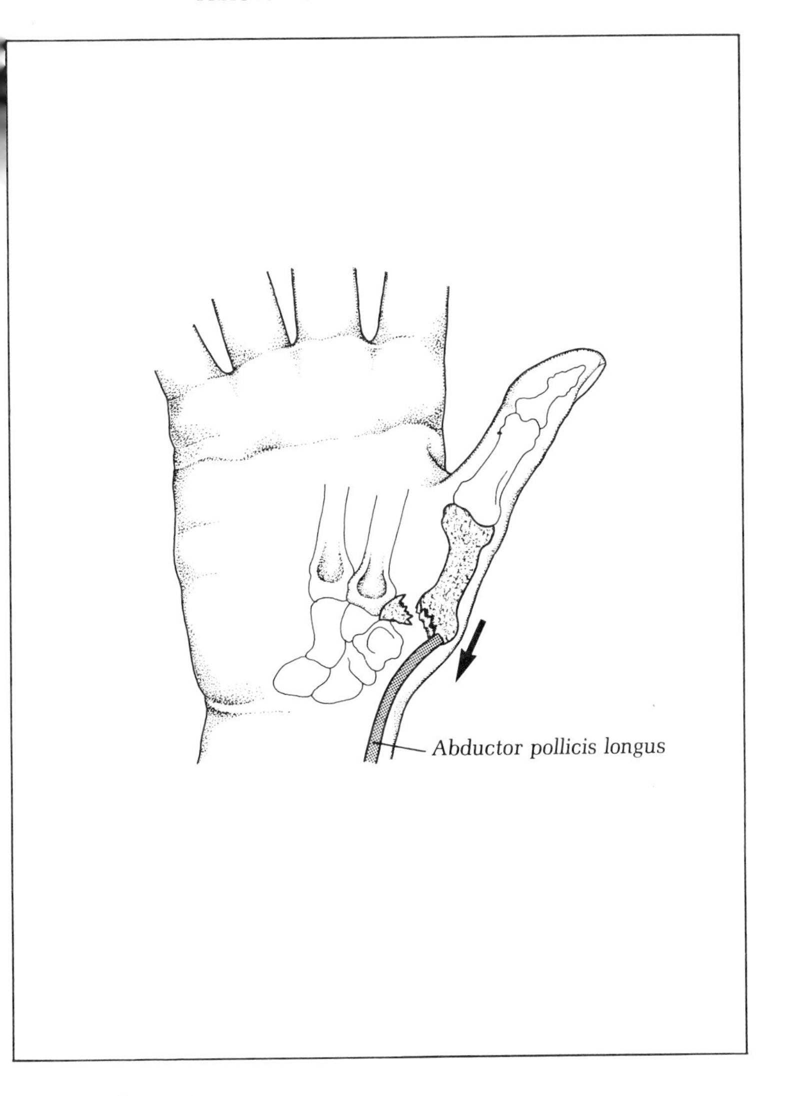

Figure 33
Bennett's fracture

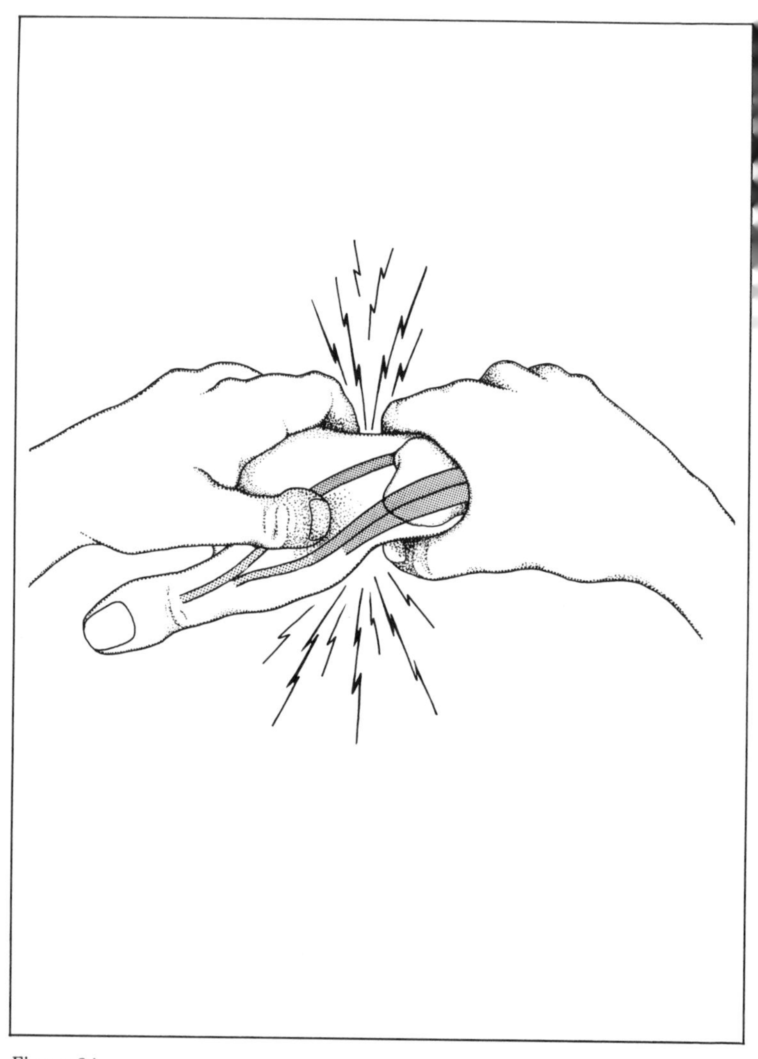

Figure 34
Fracture of scaphoid with tenderness on deep palpation in the snuff-box area

follow-up is required to make certain that these occult fractures are not overlooked.

The MCP joint dislocation

The thumb may be subjected to significant hyperextension forces. The MCP volar plate may be disrupted at it's metacarpal attachment with a hyperextension injury and the joint may dislocate so that the proximal phalanx comes to lie dorsal to the metacarpal head which buttonholes between the intrinsic muscles and the FPL.

Similarly, the finger MCP joint (most commonly index or little) may dislocate, with the metacarpal head becoming entrapped between the flexor tendon ulnarly, the lumbrical musculotendinous unit radially, and the volar plate dorsally.

These dislocations can only be detected on true lateral X-rays and almost always require open reduction.

Torn ulnar collateral ligament of thumb MCP joint

Acute radial deviation of the thumb at the MCP joint may disrupt the ulnar collateral ligament. It is commonly caused by falls while skiing, in which the thumb is forcefully radially deviated by the ski pole or strap when the hand hits the ground. It is important to compare the joint stability of the injured thumb with the patient's uninjured other thumb. The lateral stress should be applied with the MCP joint at 0° and in 35-40° of flexion. This test can be done clinically without the need for a 'stress X-ray.' If the radial deviation of the thumb on stress testing with local wrist block anesthesia is 15 degrees greater than that of the

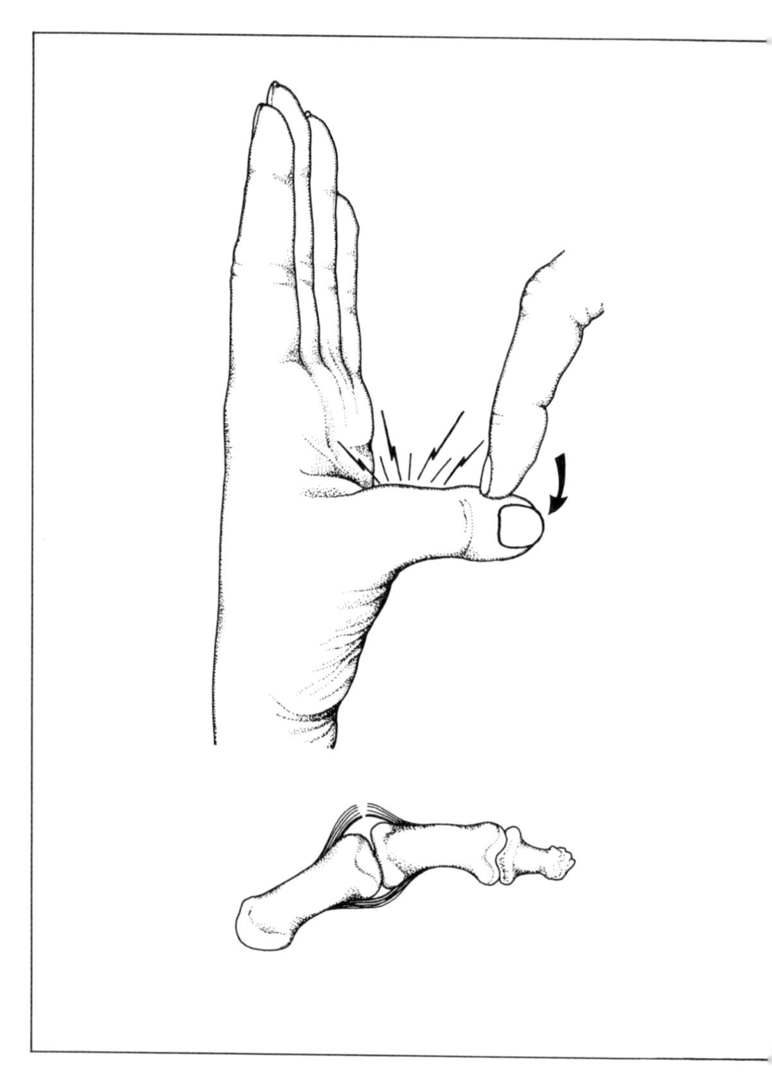

Figure 35
Rupture of ulnar collateral ligament of MCP joint of thumb

other uninjured thumb, the collateral ligament is probably disrupted. Surgical repair is usually advisable (Fig. 35).

5

ACQUIRED DEFORMITIES

Deformities of the hand may be congenital or acquired. The acquired deformities may be associated with previous traumatic injuries to joints, tendons, or nerves, with progressively contracting fascia of the palm, or with arthritis. A discussion of some of the common deformities is presented.

MALLET FINGER

The mallet finger is a flexion posture or 'droop' of the finger at the DIP joint area in which there is complete passive but incomplete active extension of the DIP joint (Fig. 36). The cause of the injury is usually a sudden blow to the tip of the extended finger. The insertion of the extensor tendon may be avulsed or there may be an avulsion fracture of the distal phalanx with a dorsal piece of bone still attached to the extensor tendon. The PIP joint should always be examined to rule out co-existing injury. Anteroposterior and true lateral roentgenograms of the PIP and DIP joints are part of the examination. A laceration over the dorsum of the distal joint may sever the extensor tendon and result in a mallet finger deformity.

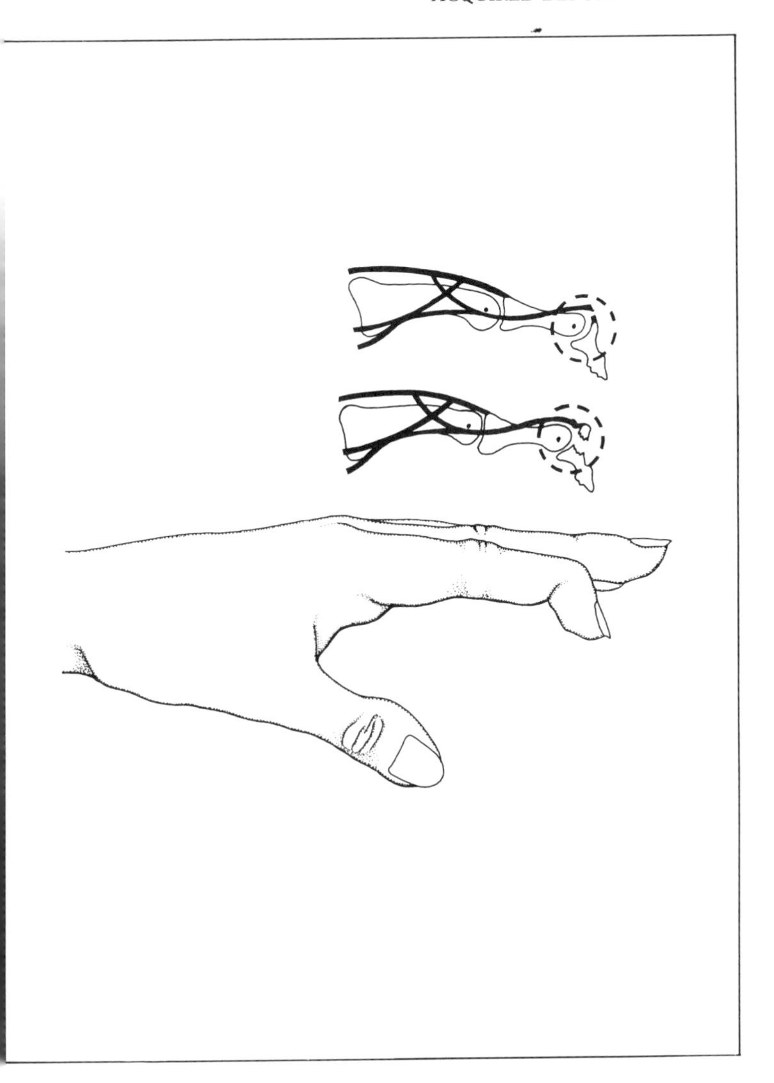

Figure 36
Mallet finger deformity (with or without fracture)

BOUTONNIÈRE DEFORMITY

In boutonnière deformity of the finger there is flexion of the PIP joint and hyperextension of the DIP joint (Fig. 37). It is the result of an injury or disease disrupting the extensor tendon insertion into the dorsal base of the middle phalanx. The fibers maintaining the position of the lateral bands progressively tear or stretch, allowing the lateral bands to slip volar to the axis of the PIP joint. They, therefore, become flexors of the PIP. The deformity may not be present, however, immediately following the injury, but can develop over several days or weeks as the lateral bands drift progressively volarward.

SWAN NECK DEFORMITY

This deformity of the finger is one in which the PIP joint is in hyperextension with the DIP joint in flexion (Fig. 38). It can be seen in a variety of conditions such as rheumatoid arthritis, certain types of spasticity, PIP joint volar plate injury, or old mallet finger deformity.

CLAW HAND

Claw hand deformity is manifest by flattening of the transverse metacarpal arch and longitudinal arches, with hyperextension of the MCP joints and flexion of the PIP and DIP joints (Figs. 25 and 39). The deformity is produced by an imbalance of the intrinsic and extrinsic muscles. The intrinsic muscles must be markedly weakened or paralyzed. The long extensor muscles

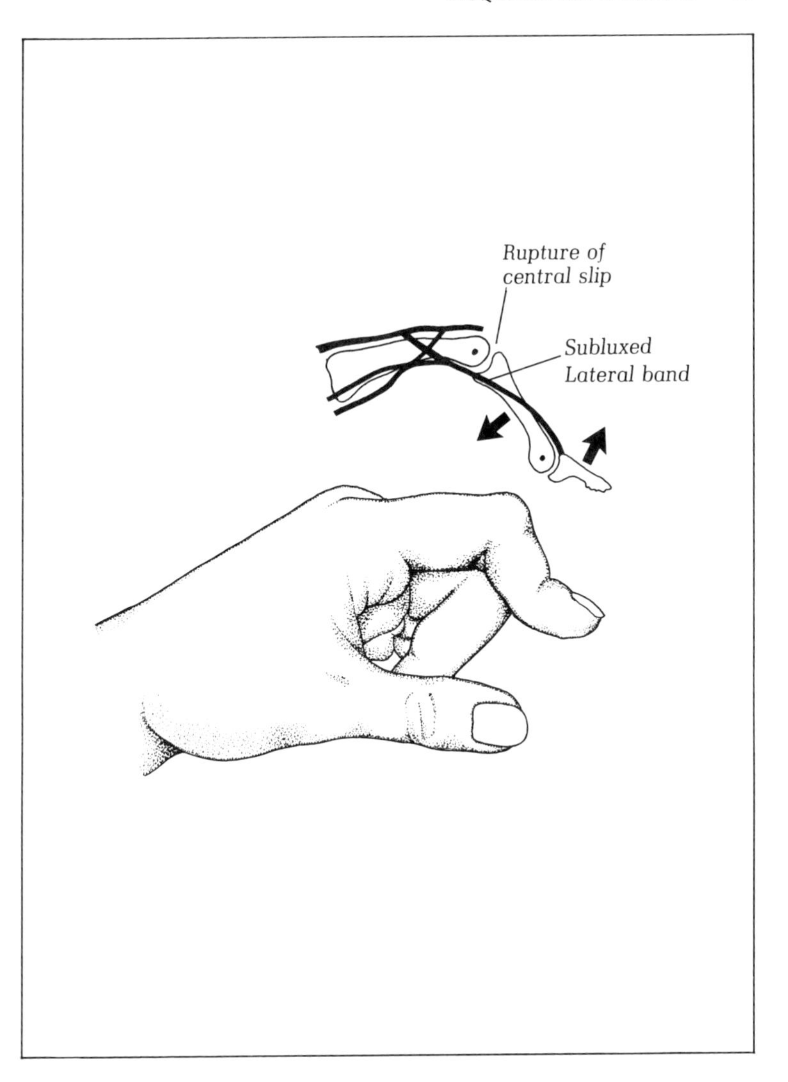

Figure 37
Boutonnière deformity

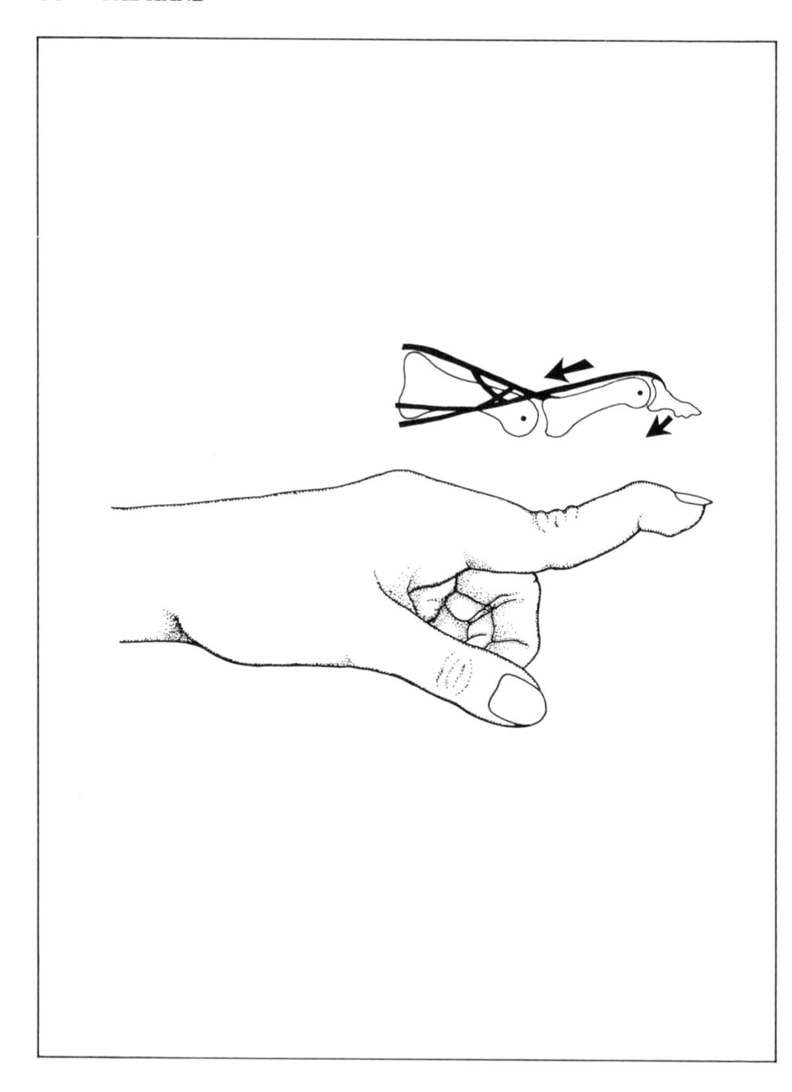

Figure 38
Swan neck deformity

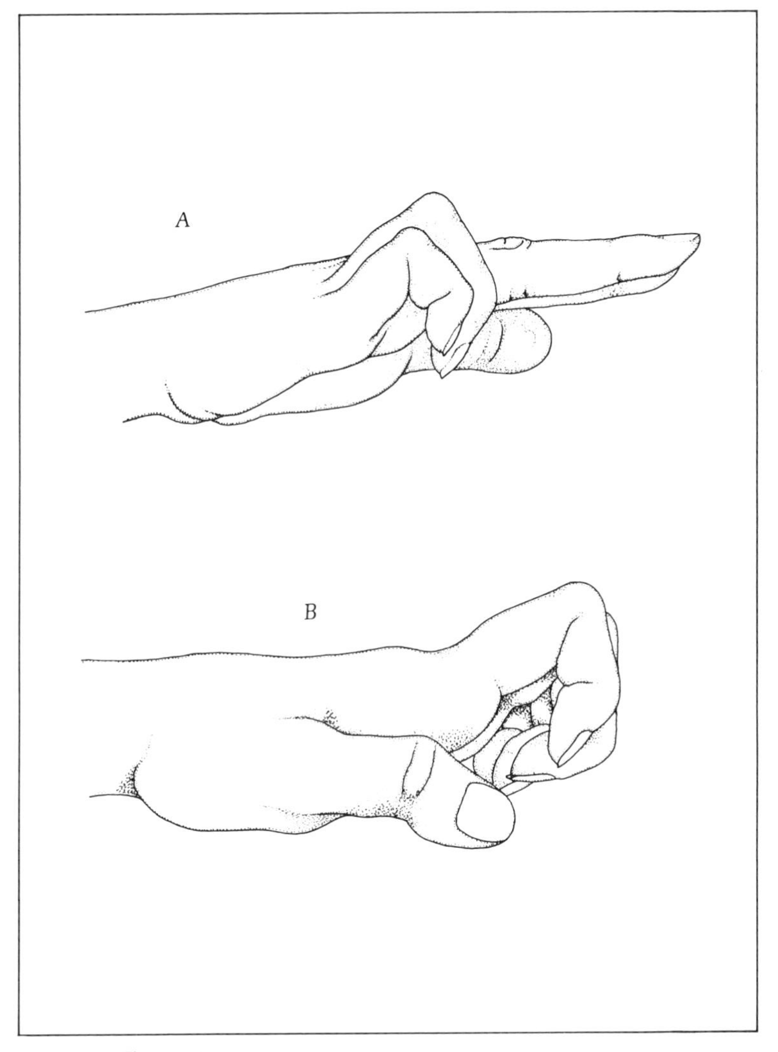

Figure 39
Claw hand deformity associated with ulnar nerve palsy (upper) and combined median and ulnar nerve palsy (lower).

hyperextend the MCP joint and the long flexor muscles flex the PIP and DIP joints. Loss of intrinsic muscle function is sometimes referred to as an 'intrinsic minus hand'. This deformity can be seen in ulnar nerve lesions, combined median and ulnar nerve lesions, brachial plexus injuries, spinal cord injuries, Charcot-Marie-Tooth disease, etc.

DUPUYTREN'S CONTRACTURE

Dupuytren's contracture is a contracture of the proliferated longitudinal bands of the palmar aponeurosis lying between the skin and flexor tendons in the distal palm and fingers (Fig. 40). The flexor tendons are not involved. It occurs most often in the ring and little fingers. It begins as a nodule and progresses to fibrous bands with contracture of the fingers. It is usually not painful and is most often seen in older men. It is often familial.

RHEUMATOID ARTHRITIS

Rheumatoid arthritis in the hand usually starts with stiff, swollen, painful fingers. The MCP and PIP joints are the ones most frequently involved. Stiffness and pain are worse on arising in the morning. As the disease progresses, the digits often become deformed and the classic ulnar drift deformity of the fingers may develop (Fig. 41). Swan neck and boutonnière deformities are common. Carpal tunnel syndrome, trigger finger, wrist tenosynovitis, painful flexor tenosynovitis, and rupture of tendons may be present.

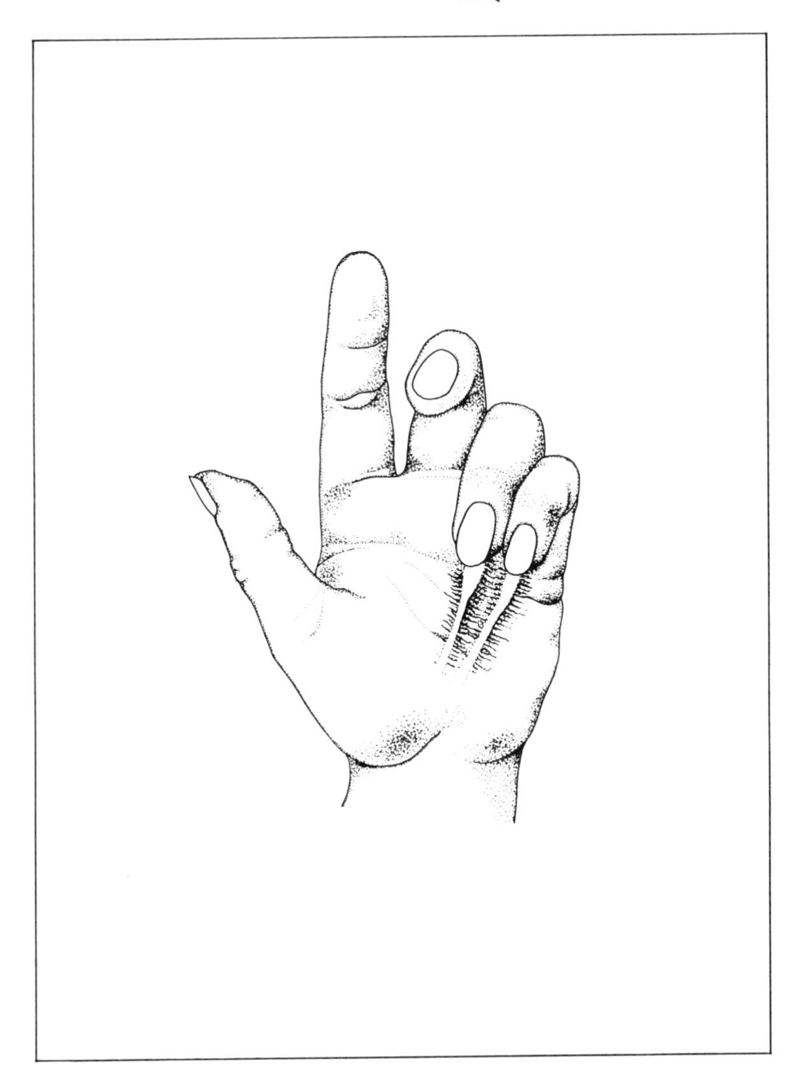

Figure 40
Dupuytren's contracture

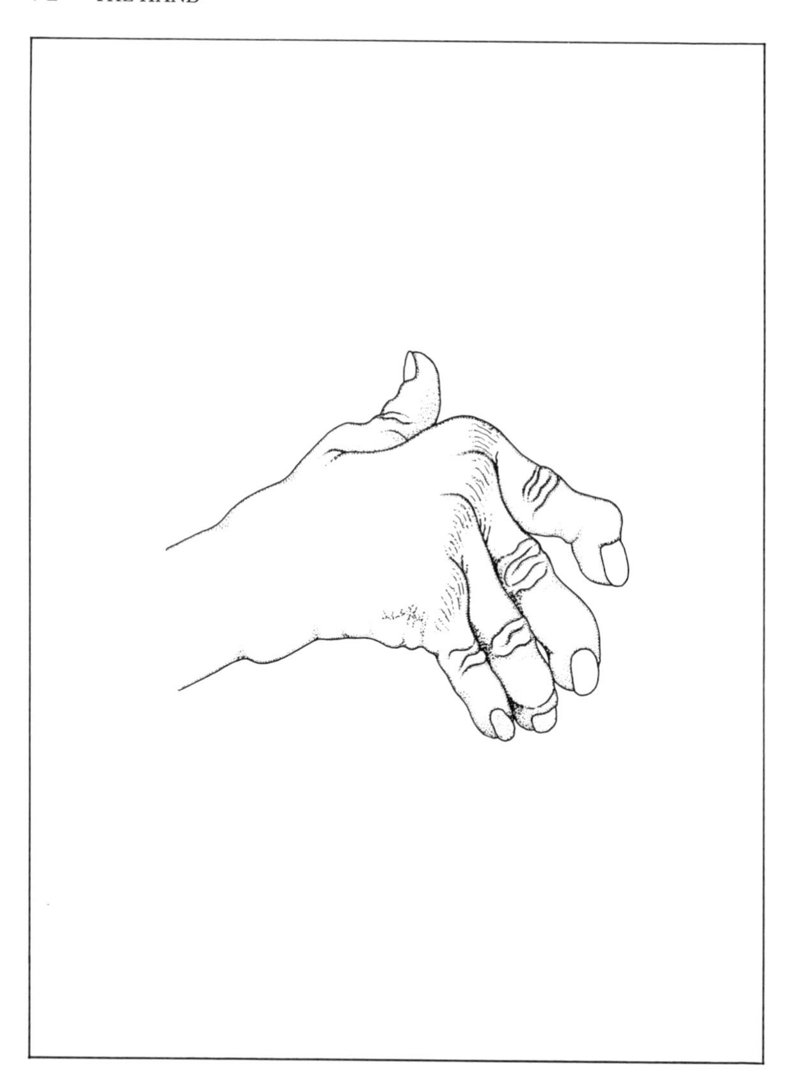

Figure 41
Rheumatoid arthritis of the hand with ulnar drift

DEGENERATIVE ARTHRITIS

In degenerative arthritis (osteoarthritis) of the hand, the distal joints develop marginal osteophytes known as Heberden's nodes (Fig. 42). Similar bony lesions may occur at the PIP joint, called Bouchard's nodes. The MCP joints are seldom involved.

The carpometacarpal joint of the thumb is also a common site for degenerative arthritis in the hand. The axial compression-adduction test (Fig. 43) is done by manipulating the thumb with axial compression and gentle adduction. The instability and crepitus are appreciated with the examining thumb placed on the joint and base of metacarpal. This is usually painful for the patient with joint involvement. The axial compression and rotation (Fig. 43) are often painful when the proximal phalanx is used as a lever arm for a grinding maneuver of the thumb carpometacarpal joint.

DEQUERVAIN'S TENOSYNOVITIS

A nonspecific tenosynovitis of the APL and EPB tendons in the first dorsal wrist compartment is known as deQuervain's disorder. Tenderness and crepitation may be present over the radial styloid. Finkelstein's test may be positive (Fig. 44). It is performed by having the patient grasp the thumb with the fingers (thumb in palm) and ulnar deviate the wrist. If this causes pain the test is positive.

It is important to differentiate between deQuervain's tenosynovitis and carpometacarpal joint arthrosis of the thumb. To do this, tenderness and pain must be accurately localized between this first extensor compartment and the carpometacarpal joint of the

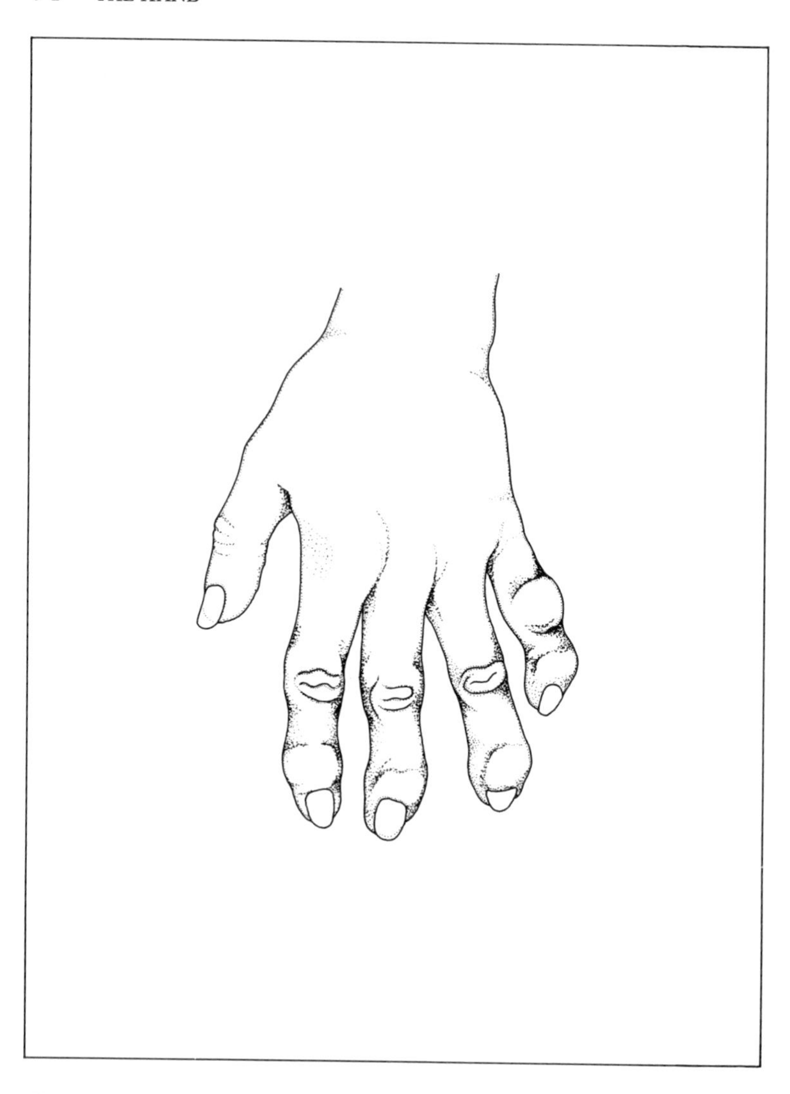

Figure 42
Degenerative arthritis of the hand

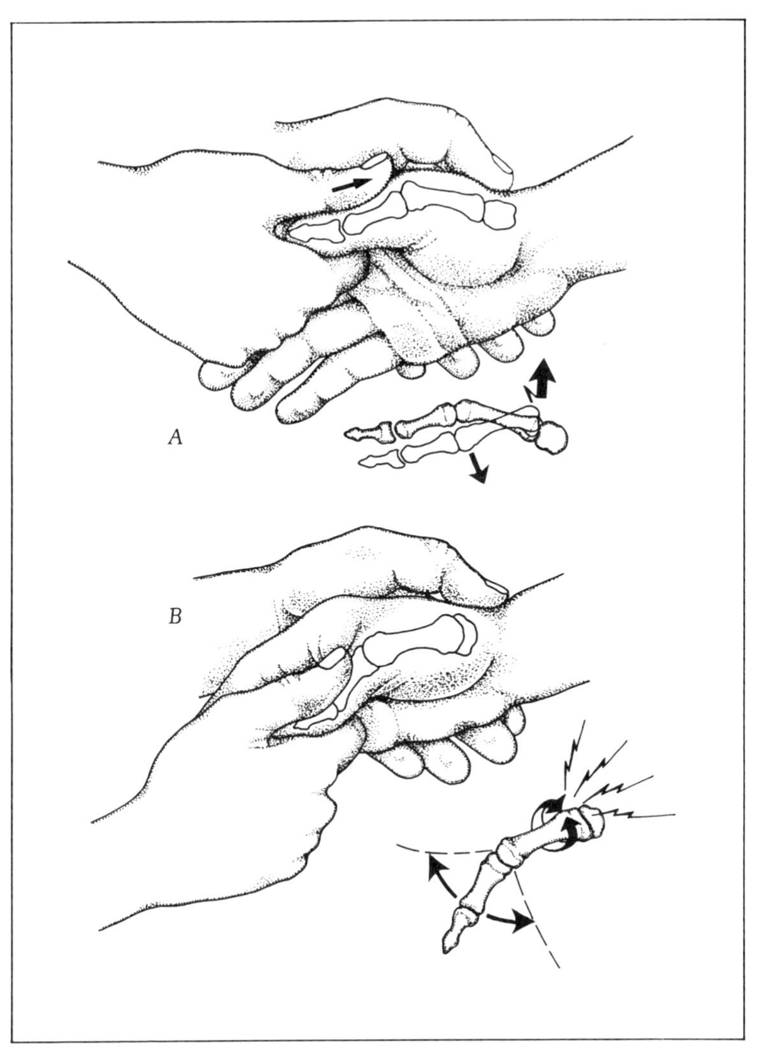

Figure 43
Upper: axial compression-adduction test. Lower: axial compression and rotation test

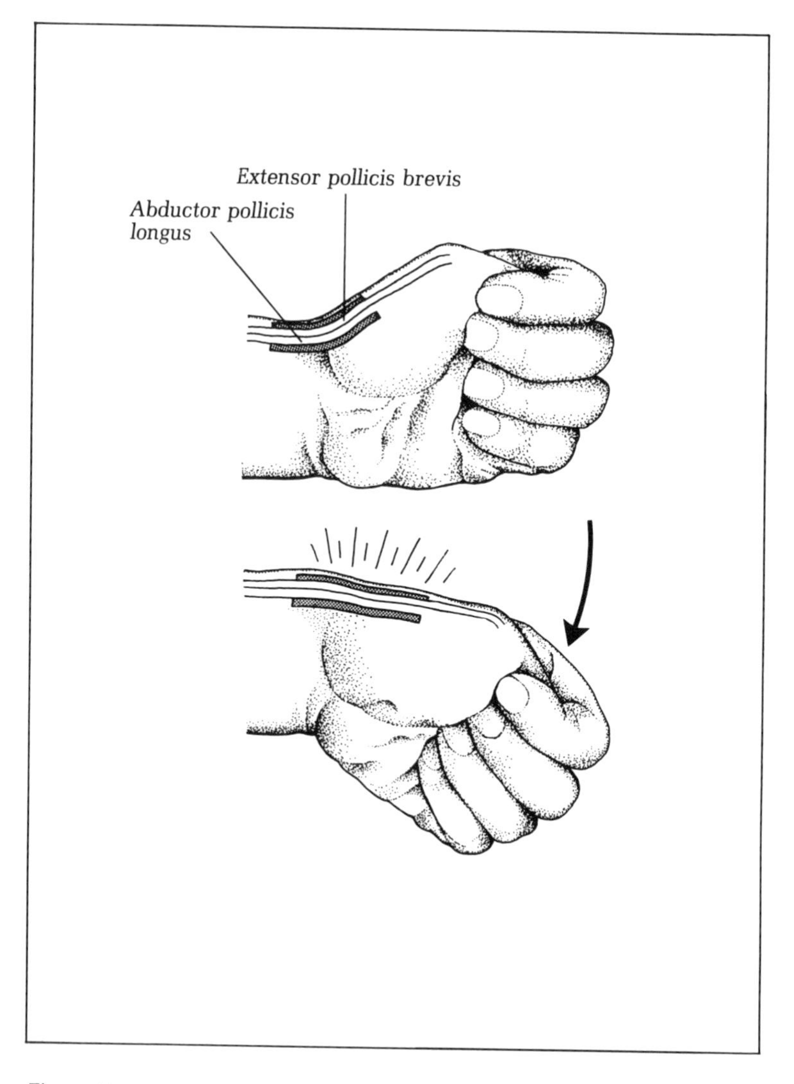

Figure 44
Finkelstein's test for deQuervain's disease

thumb. Roentgenograms can be obtained if there is uncertainty.

TRIGGER THUMB AND TRIGGER FINGER

Stenosing tenosynovitis can occur in the thumb or any finger but it most commonly occurs in the ring or middle fingers. Inflammation at the MCP joint pulley causes a discrepancy between the size of the tendon and pulley. The tendon may become thickened just proximal to the pulley. This discrepancy in size may cause a snapping or locking phenomenon, holding the thumb or finger flexed or extended (Fig. 45). Palpation of the flexor tendon over the MCP joint can be painful.

CARPAL TUNNEL SYNDROME

The carpal tunnel syndrome is a median nerve compression neuropathy at the wrist where the nerve passes beneath the transverse carpal ligament (Fig. 46). Patients complain of their hands 'going to sleep' and are frequently awakened at night with numbness, pain, and tingling in the thumb, index, long, and ring fingers. The little finger is not usually involved. Patients may complain of referred pain in the forearm and even as high as the shoulder. They notice these symptoms when driving a car or during other sustained activities. The entity is more common in women than men. The dominant hand is more often involved but symptoms can be bilateral. The carpal tunnel syndrome may be associated with rheumatoid arthritis or following a Colles' fracture. It can also be seen in a variety of medical conditions such as pregnancy, diabetes, and

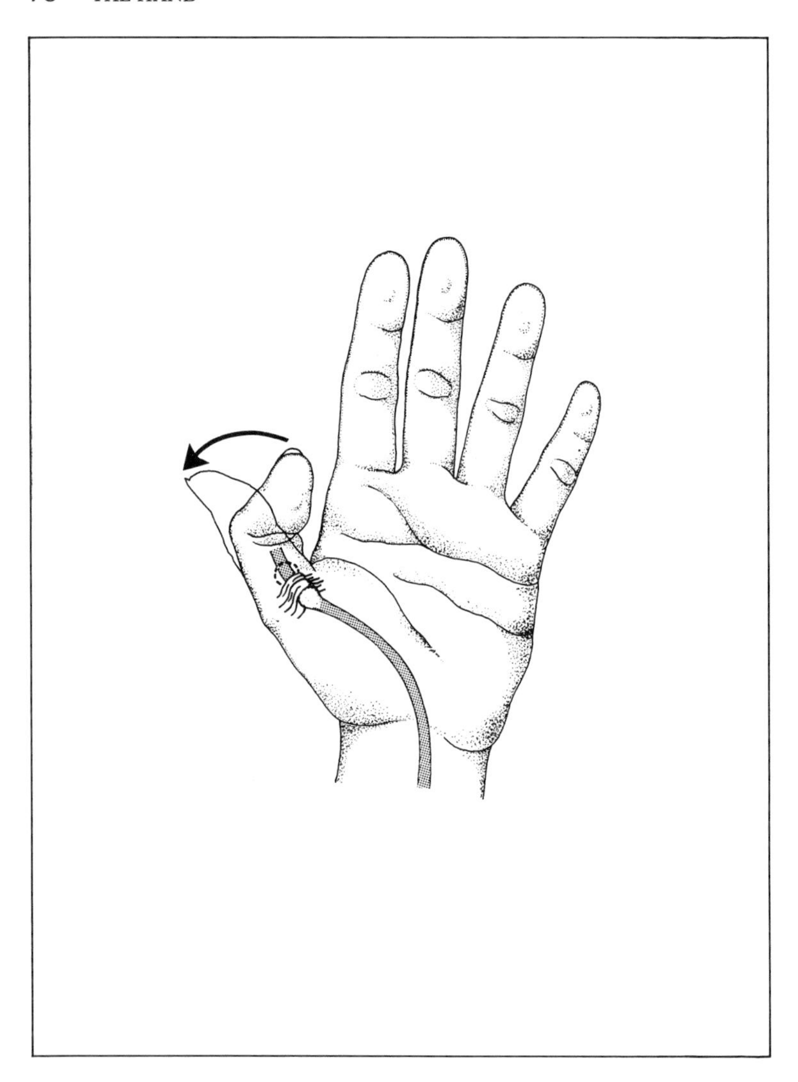

Figure 45
Trigger thumb

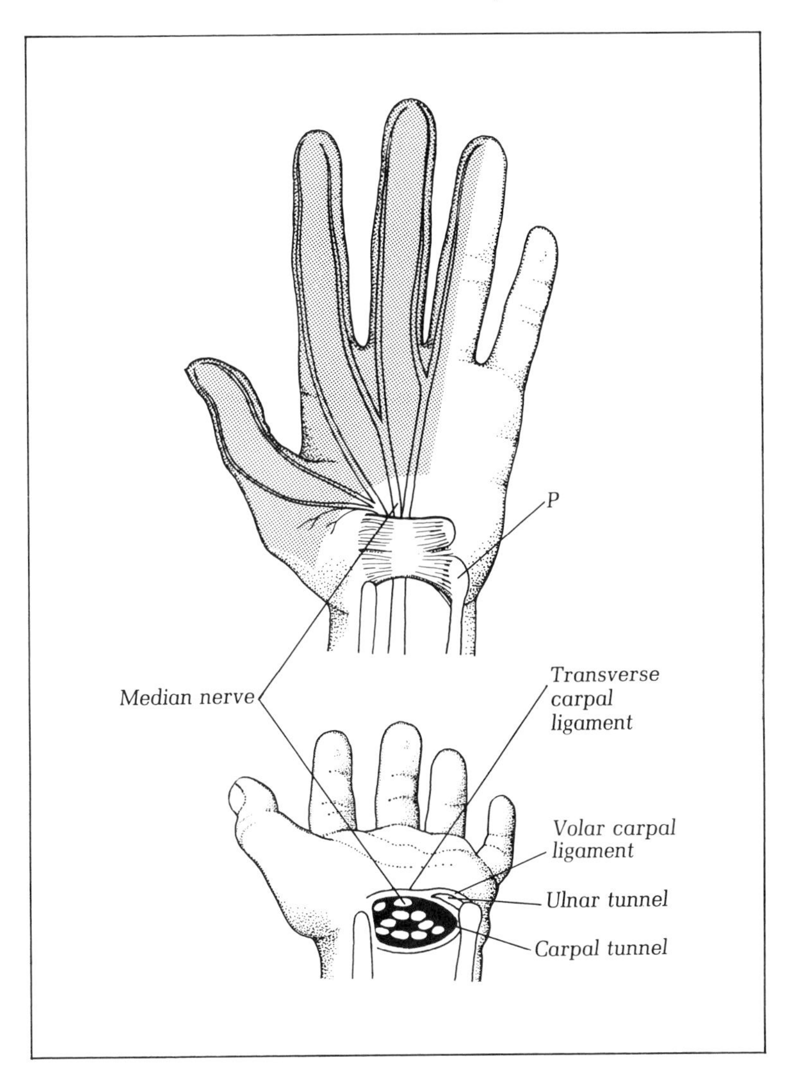

Figure 46
Carpal tunnel syndrome (median nerve compression syndrome)

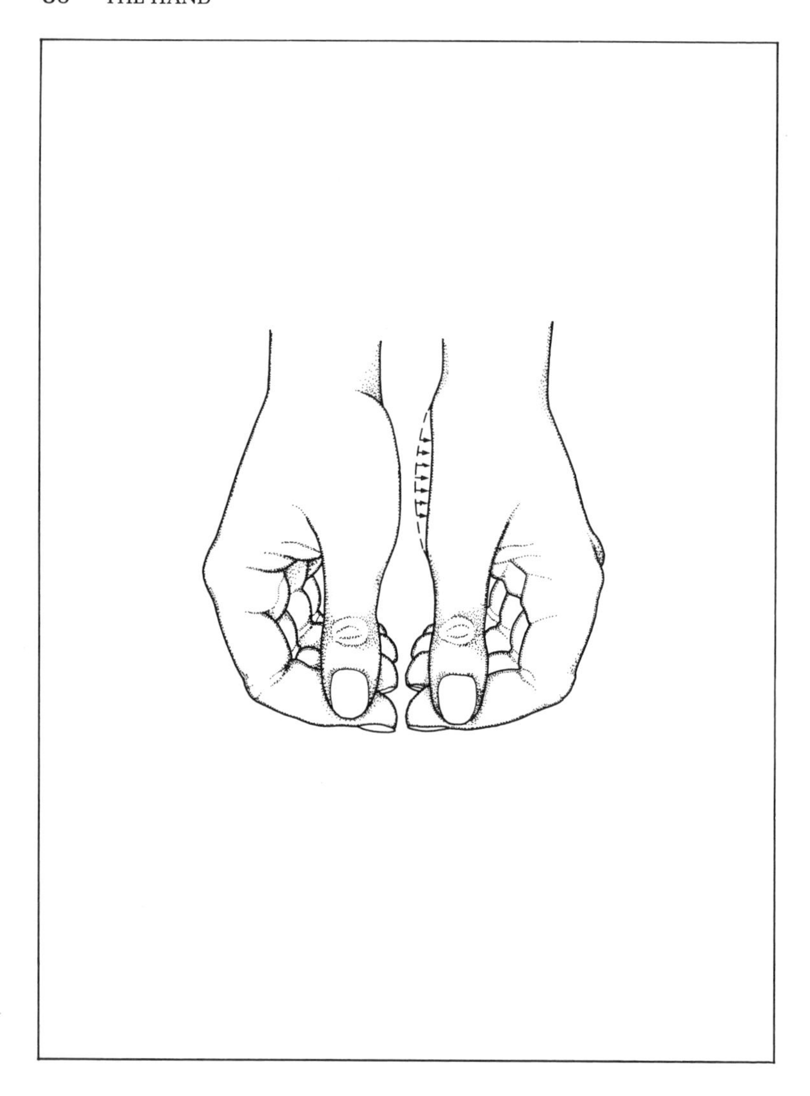

Figure 47
Atrophy of the thenar muscle

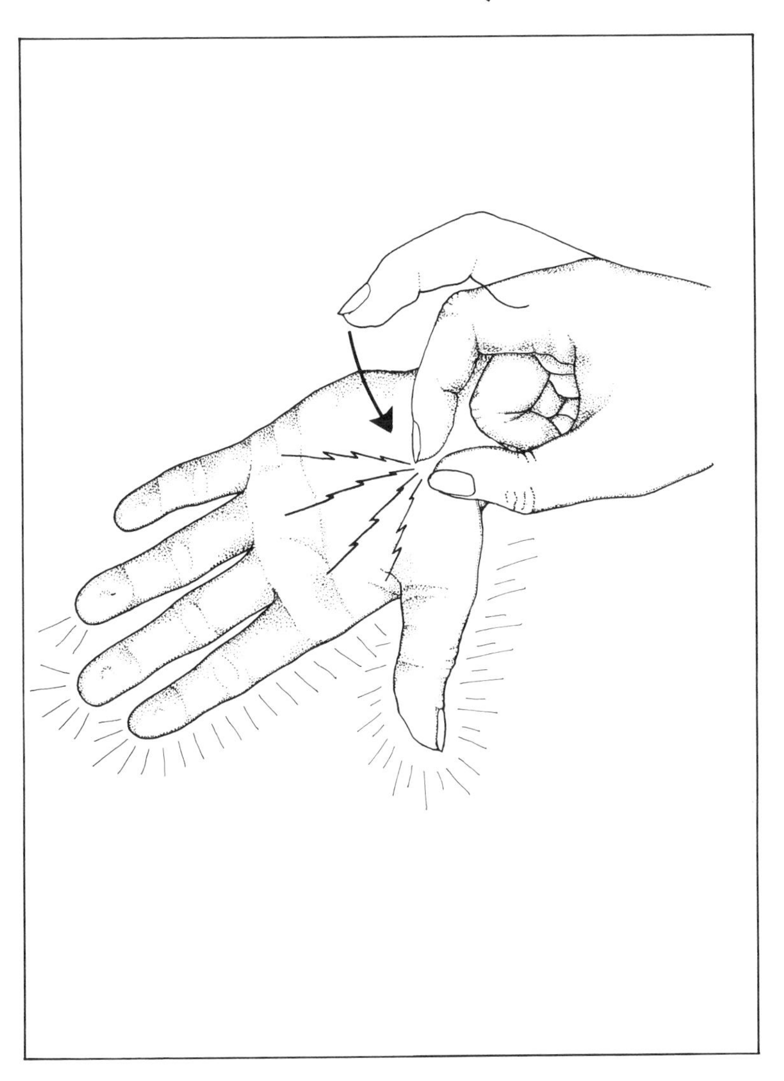

Figure 48
Tinel's sign

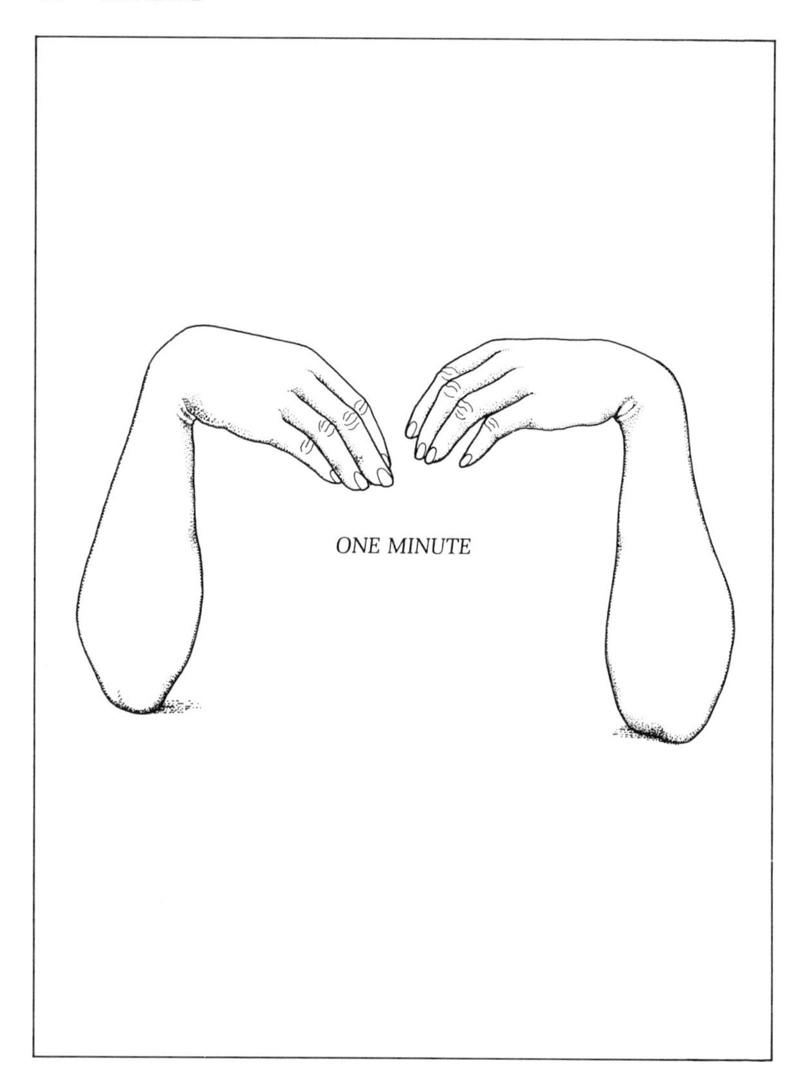

Figure 49
Phalen's sign (wrist flexion test)

thyroid disease. However, most patients with the carpal tunnel syndrome have no apparent associated systemic disease.

The hand will most often look normal; however, in long-standing cases there may be atrophy of the median innervated thenar muscles (Fig. 47).

Tapping over the median nerve at the wrist crease may produce paresthesias in the hand (Tinel's sign) (Fig. 48). The wrist flexion test (Phalen's test) (Fig. 49) is done by resting the elbows on a table and allowing the wrists to fall into complete volar flexion for one minute. If the patient has carpal tunnel syndrome, this position may produce paresthesias in the hand. The tourniquet test is done by inflating the blood pressure cuff above 200 mmHg for two minutes; pain, numbness, and tingling may develop in the median nerve distribution area.

6

CONGENITAL ANOMALIES

Congenital defects are often encountered in the examination of the hand. These defects should be recorded in an accurate and complete manner. In the past, the use of various Greek and Latin names to describe common deficiencies has only served to confuse many clinicians. A classification should be used that groups cases according to the parts that have been affected primarily by certain embryological failures.

The various clinical pictures of limb deficiencies are felt to represent varying degrees of destruction within the ectomesenchymal mass that develops on the lateral body wall of the developing embryo. The limb bud is first noted at the fourth week after gestation. These buds grown and differentiate rapidly in a proximodistal sequence during the following four weeks. Any factor, environmental or otherwise, that disrupts the sequential differentiation during this period will produce a defect in the limb compatible with the timing of the insult.

Congenital defects should be classified in the following categories as outlined by the American Society for Surgery of the Hand and the International Federation of Societies for Surgery of the Hand:

1. Failure of formation of parts (arrest of development).
2. Failure of differentiation (separation) of parts.
3. Duplication.
4. Overgrowth (gigantism).
5. Undergrowth (hypoplasia).
6. Congenital constriction band syndrome.
7. Generalized skeletal abnormalities.

FAILURE OF FORMATION OF PARTS

The category of failure of formation of parts is that group of congenital deficiencies noted by failure or arrest of formation of the limb either complete or partial. This category is divided into two types: *transverse* and *longitudinal*.

Transverse deficiencies

Transverse deficiencies represent the so-called congenital amputations ranging from aphalangia (absence of the fingers) to amelia (absence of the extremity). The stumps are usually well padded and may show rudimentary digits or dimpling. One of the most common defects in the group is the short below elbow amputation. This would be classified as a transverse (T), left or right (L-R), forearm (FO), upper one-third deficiency.

Longitudinal deficiencies

Longitudinal deficiencies include all other limb deficiencies in this category. In identifying longitudinal deficiencies, all completely or partially absent bones are named. The deficiencies in the group reflect the separation of the pre-axial (radial) and post-axial (ulnar) divisions in the limb and include longitudinal failure of formation of the entire limb segment (phocomelias) or either radial, central, or ulnar components of the limb.

An example of a segmental failure would be the phocomelia (hand attached to the trunk). This would be classified: longitudinal (L), left or right (L-R), humerus (Hu), radius (Ra), ulna (Ul).

The absence of parts of the radial (pre-axial) side of the limb may vary from deficient thenar muscles to a short floating thumb, and from deficient carpals, metacarpals, and radius to the classified so-called radial club hand. The classification of longitudinal, right, radius, proximal one-third, carpal partial, 1st ray, would be a deficiency with a partial absence of the radius and carpal bones with no thumb on the right.

Central deficiencies include deficiencies of the middle three digits: index, long, and ring, and sometimes the carpal bones. The middle digit may be missing in the so-called lobster claw hand.

In ulnar deficiencies, the little or ring finger may be missing and can be associated with partial or complete absence of the ulna and carpal bones. These are classified in a like manner.

FAILURE OF DIFFERENTIATION (SEPARATION) OF PARTS

Failure of differentiation is that category in which the basic units have developed but the final form is not completed. The homogenous anlage divides into separate tissues of skeletal, dermomyofascial or neurovascular elements found in normal limbs, but fails to differentiate completely or to separate. An example in the forearm would be a synostosis (fusion of bones which are normally separated) of the proximal radius and ulna. In the wrist, fusion of carpal bones is frequently seen as well as fusion of two or more metacarpals. Synphalangism is end to end fusion of the proximal interphalangeal joints. Syndactyly is by far the most common deformity seen in this category. The failure of differentiation can vary from simple skin bridging to fusion of parts.

Contractures secondary to failure of differentiation of muscle, ligaments, and capsular structure are frequently seen. They vary from simple trigger thumb to flexion contractures of the little finger (camptodactyly) to the severe arthrogryposis of the hand.

Lateral deviation or displacement due to asymmetrical abnormalities of the digits (clinodactyly) also occur.

DUPLICATION

Duplication of parts probably occurs as a result of a particular insult to the limb bud and ectodermal cap at a very early stage of their development so that splitting of the original embryonic part occurs. These defects may range from polydactyly (too many digits) to twinning or mirror hand (duplication of the digits present). They are classified according to the parts or tissues duplicated. Polydactyly is the most common deformity seen in this group. It can be either radial (duplication of the thumb, partial or complete), central (middle three fingers) or ulnar (little finger duplication, partial or complete). The thumb and little finger duplications are seen more frequently.

OVERGROWTH (GIGANTISM)

In this category, there can be overgrowth of the entire limb or a single part. Some cases appear to be due to skeletal overgrowth with normal-appearing soft tissue. Others show excess fat, lymphatic, and fibrous tissue; neurofibromata, lymphangiomata, or angiomata may be present in these cases. A frequently seen deformity in this category is gigantism of the digit and there can be an accompanying syndactyly. This would be classified as an overgrowth (gigantism) of the digit with syndactyly as a secondary condition.

UNDERGROWTH (HYPOPLASIA)

Undergrowth or hypoplasia denotes defective or incomplete development of the parts. This may be manifested in the entire extremity or its divisions. Hypoplasia may involve any of the following systems: skin and nails, musculotendinous, neurovascular, or the extremity (arm, forearm, hand). An abnormally short, completely formed metacarpal would be brachymetacarpia. Brachyphalangia refers to abnormally short middle phalanges.

7

TUMORS

The most common soft tissue mass of the hand is a ganglion (Fig. 50). It has a well-defined, smooth surface and is a firm cystic lesion that is fixed to the deep tissues. It may develop over the volar or dorsal area of the wrist, originating from the wrist capsule. Those in the palm near the digital palmar skin crease arise from the flexor tendon sheath and may or may not be painful.

A mucous cyst is a cystic lesion (actually a ganglion) over the dorsum of the finger near the distal joint and fingernail (Fig. 51). It is associated with degenerative arthritis of the DIP joint of the finger and arises from the joint. It may have thin walls and there may be associated grooving of the fingernail distal to the cyst. Should a mucous cyst rupture and become infected, a septic joint may result.

Other soft tissue tumors of the hand which may present as a mass are giant cell tumors of tendon sheath, pigmented villonodular synovitis, and inclusion cysts.

Malignant tumors of the soft tissue and bone are rare in the hand. However, skin cancers (basal cell and squamous cell) on the dorsum of the hand are seen in the elderly. Malignant melanoma does occur in the hand and may be subungual.

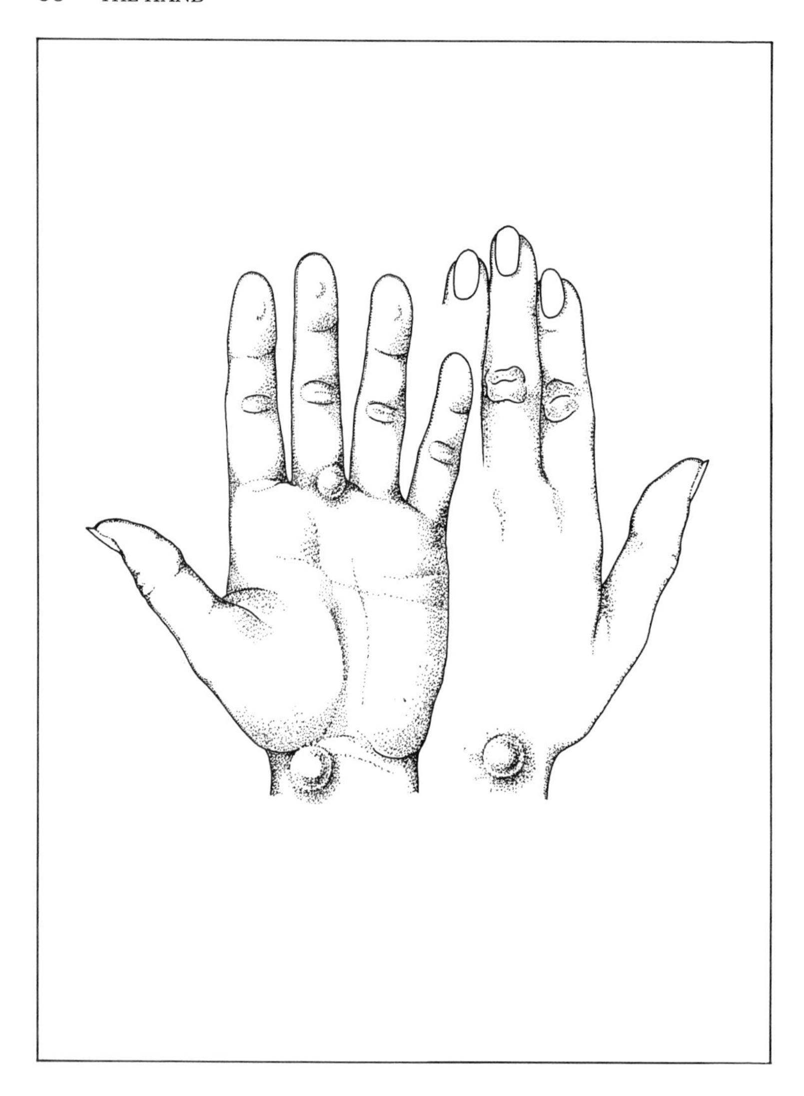

Figure 50
Ganglion of the hand

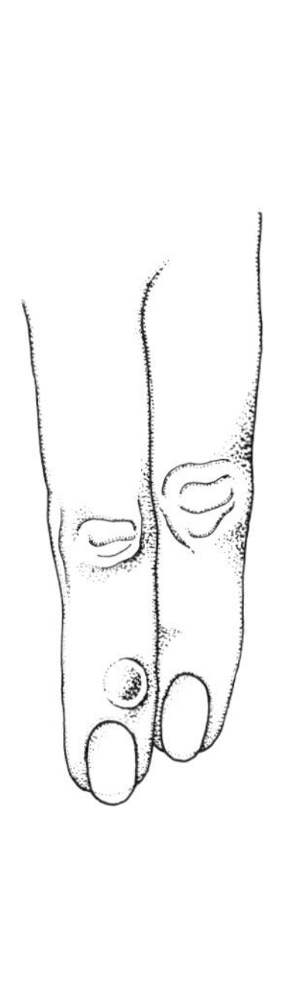

Figure 51
Mucous cyst

Primary bone tumors of the hand usually present as swelling and/or pain in the area of the hand involved. The tumor is located with a roentgenogram of the hand and the diagnosis is established by biopsy of the tumor.

8

INFECTION

PARONYCHIA

A paronychia is an infection of the soft tissue around the fingernail which usually begins as a ‘hangnail’ and which is usually caused by a staphylococcus infection (Fig. 52A). It spreads around the nail eponychium, thus the term ‘run around’. It is red, swollen, and very painful, with purulent drainage around the margin of the nail.

FELON

A felon is a deep infection of the pulp space of the distal segment of the finger (Fig. 52B). The distal segment is swollen, red, and extremely painful. Drainage is usually required. It is usually caused by a staphylococcus infection and can involve the distal phalanx with osteomyelitis.

PURULENT TENOSYNOVITIS

Infection of the tendon sheath of the digit presents as a (1) swollen, (2) slightly flexed finger with (3) tenderness

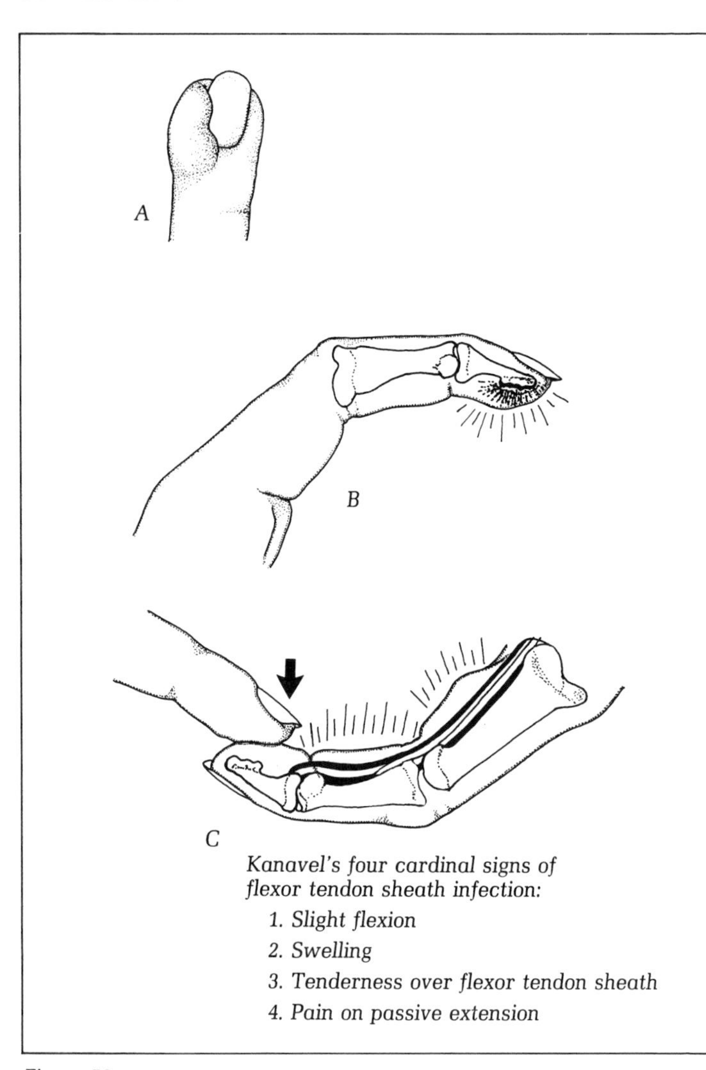

Figure 52
a. Paronychia
b. Felon
c. Flexor tendon sheath infection

over the flexor tendon sheath and (4) increased pain on passive extension of the digit (Fig. 52C). These findings constitute Kanavel's four cardinal signs of a purulent tendon sheath infection.

If the tendon sheath of the little finger or thumb is involved primarily, the infection may spread to the wrist area where the sheaths communicate and the classic 'horseshoe' infection may develop (Fig. 53). The sheath of the index, long, and ring fingers extends to the palm but not to the wrist. Streptococcus and staphylococcus are the most frequent infecting organisms.

These are serious infections which may extend along the flexor tendon sheath and prompt treatment is most important.

SPACE INFECTIONS

Thenar space and mid-palm infections are not common. When they do occur the dorsum may be more swollen than the palm of the hand. This should not mislead the examiner. The usual findings of redness, tenderness, and perhaps fluctuance help to define the abscess.

The thenar space is a potential space anterior to the adductor muscle (Fig. 54). Its ulnar border is separated from the mid-palm space by a fascia arising from the metacarpal of the long finger and attaching to the palmar fascia.

The mid-palm space is the potential space anterior to the interosseous muscles and posterior to the flexor tendons of the long, ring, and little fingers (Fig. 55).

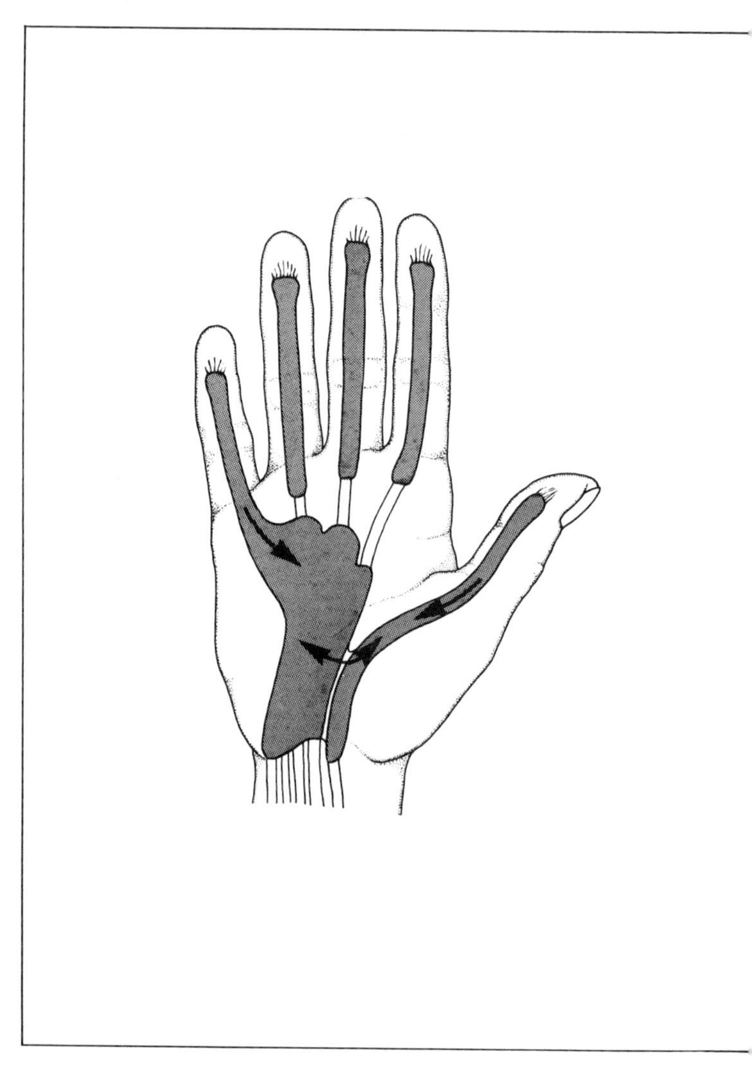

Figure 53
Tendon sheaths of the flexor tendon

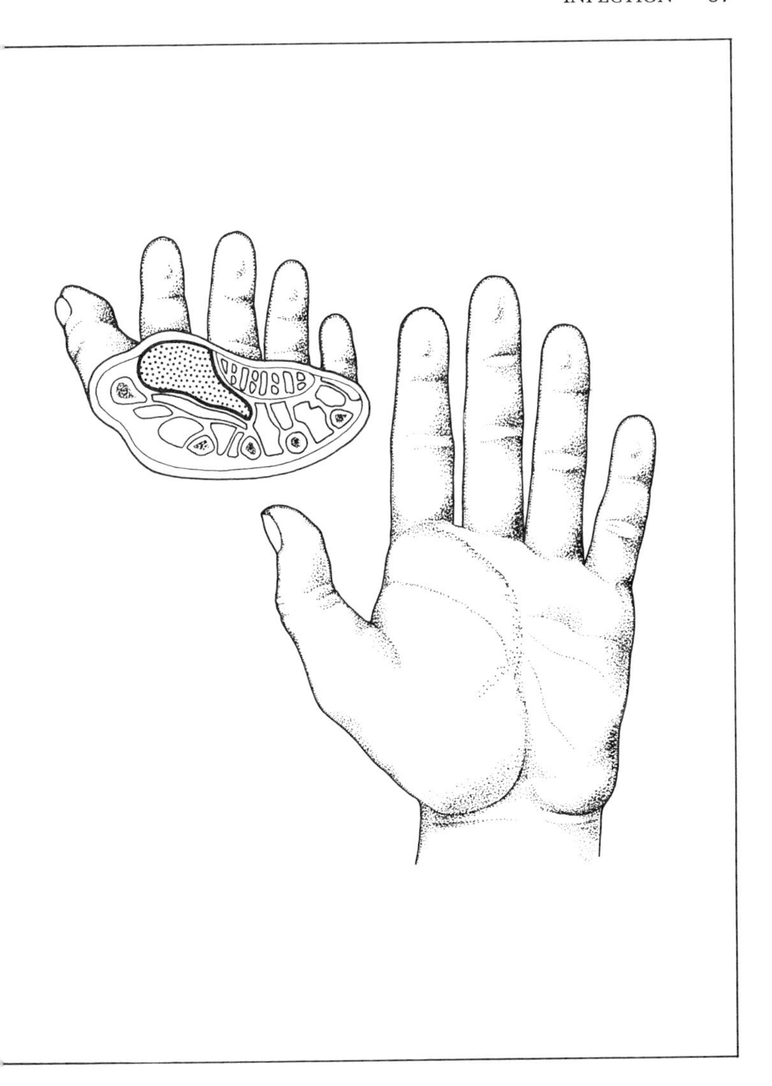

Figure 54
Thenar space infection of the hand

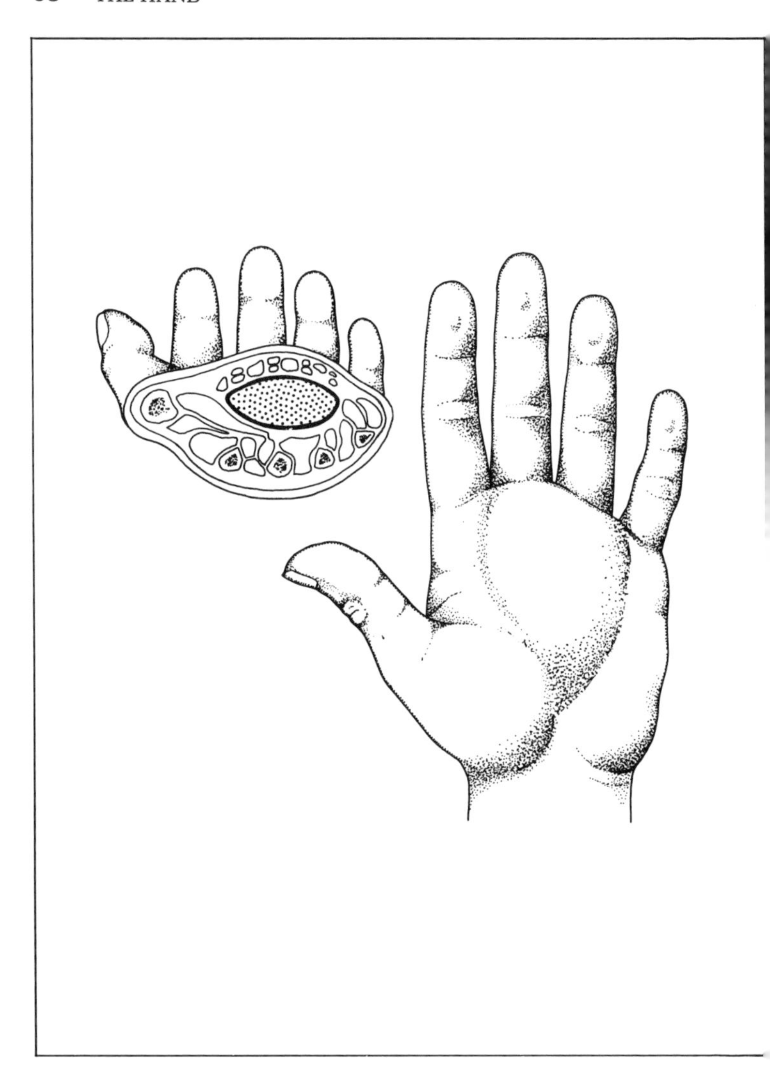

Figure 55
Mid-palm space infection of the hand

HUMAN BITE INFECTIONS

Human bites are commonly seen over the dorsum of the MCP joints. These usually occur when the joint area strikes a tooth in a fight. The important point is that the wound may appear benign initially, but is usually inoculated with a potent mixture of bacterial flora. This is a serious injury requiring prompt treatment.

APPENDIX 1

KEY TO ABBREVIATIONS USED IN TEXT

ADM	Abductor digiti minimi
AdP	Adductor pollicis
APB	Abductor pollicis brevis
APL	Abductor pollicis longus
CMC	Carpometacarpal
DIP	Distal interphalangeal
ECRB	Extensor carpi radialis brevis
ECRL	Extensor carpi radialis longus
ECU	Extensor carpi ulnaris
EDC	Extensor digitorum communis
EDM	Extensor digiti minimi
EIP	Extensor indicis proprius
EPB	Extensor pollicis brevis
EPL	Extensor pollicis longus
FCR	Flexor carpi radialis
FCU	Flexor carpi ulnaris
FDM	Flexor digiti minimi
FDP	Flexor digitorum profundus
FDS	Flexor digitorum superficialis
FPB	Flexor pollicis brevis
FPL	Flexor pollicis longus
I	Index finger

IP	Interphalangeal
L	Little finger
M	Middle finger
MCP	Metacarpophalangeal
ODM	Opponens digiti minimi
OP	Opponens pollicis
PIP	Proximal interphalangeal
PL	Palmaris longus
R	Ring finger

APPENDIX 2

ANATOMY—SUMMARY

Joint Control	Prime* Muscle	Nerve	Fig. No.†	Comments
Wrist				
Flexion	FCR PL FCU	Median Median Ulnar		Absence — weak wrist flexion present by FDS, FDP
Extension	ECRL ECRB ECU	Radial	7, 9, 12	Absence — 'wrist drop'
Radial deviation	ECRL FCR	Radial Median		
Ulnar deviation	ECU FCU	Radial Ulnar		
Finger MCP				
Flexion	Interosseous Lumbrical	Ulnar Median I. & M.; Ulnar R. & L.	39	Absence — claw hand
Extension	EDC EIP EDM	Radial	7, 11, 29	Absence — MCP extensor lag
Abduction	Dorsal Interosseous	Ulnar	15, 16	
Adduction	Volar Interosseous	Ulnar		

Joint Control	Prime* Muscle	Nerve	Fig. No.†	Comments
Finger PIP				
Flexion	FDS	Median	6	Must block FDP to detect clinical absence
Extension	Interosseous	Ulnar		Intrinsic independent of MCP position: extrinsic only if MCP joint flexed or at 0° (i.e., not hyperextended)
	Lumbrical	Median & Ulnar	39	
	EDC, EIP, EDM	Radial		
Finger DIP				
Flexion	FDP	Median, I. & M.; Ulnar R. & L.	5	
Extension	None	—		Strong DIP extension contingent upon active PIP extension control
Thumb CMC				
Flexion-adduction	AdP	Ulnar		
	Ulnar ½ FPB	Ulnar or Median	14	
	1st dorsal interosseous	Ulnar		
	FPL	Median		
Extension-abduction	EPL, EPB	Radial	8, 10	
	APL	Radial		
	APB	Median		

Thumb CMC (cont'd)				
Opposition (Pronation)	APB Radial ½ FPB OP	Median	3B, 13	A composite motion
Supination	EPL	Radial		
Thumb MCP				
Flexion	FPL thenar intrinsic muscles (except OP)	Median Median Ulnar		
Extension	EPB	Radial	8	
Thumb IP				
Flexion	FPL	Median	4	
Extension	EPL	Radial	10	Weak IP extension also by intrinsics

(*achieves a given function but does not imply 'the strongest' acting across that joint)
(†in addition to Figs. 18, 19, 21)

APPENDIX 3

CLINICAL ASSESSMENT RECOMMENDATIONS

SENSIBILITY

Two-point discrimination with the use of a blunt instrument, applied in a longitudinal axis of the digit (Fig. 23). Pressure applied which does not blanch skin.

Ratings

1. Normal, less than 6 mm.
2. Fair, 6 to 10 mm.
3. Poor, 11 to 15 mm.
4. Protective, one point perceived.
5. Anesthetic, no point perceived.

STRENGTH

Grip strength

Use a squeeze (grip) dynamometer and make three successive determinations: record, and calculate percentage relative to pre-treatment value as well as to value from contralateral hand. Note: this is *NOT* a percentage of physical impairment or improvement but merely an indicator of improving or worsening condition.

Pinch strength

Use a pinch dynamometer. Key pinch is the thumb tip to radial aspect of middle phalanx of index finger and is the most universal and preferred value. Record three successive efforts and calculate percentage relative to pre-treatment as well as contralateral hand values. Tip pinch value (reverse key pinch — index tip to ulnar tip of thumb) will be less powerful than key pinch. Same recordings as for key pinch. Note: this is *not* a percentage of physical impairment or improvement but merely an indicator of improving or worsening condition.

MOTION

Total passive motion (TPM)

Sum of angles formed by MCP, PIP and DIP joints in maximum passive flexion minus the sum of angles of deficit from complete extension at each of these three joints: (MCP + PIP + DIP) − (MCP + PIP + DIP) = total flexion − total extensor lag = TPM.

Total active motion (TAM)

Sum of angles formed by MCP, PIP, and DIP joints in maximum active flexion, i.e., fist position, minus total extension deficity at the MCP, PIP, and DIP joints with active finger extension. Significant hyperextension at any joint, particularly the PIP and DIP joints, is recorded as a deficit in extension and is included in the total extension deficit. Hyperextension must be considered an abnormal value in swan neck, (PIP) and boutonnière deformities (DIP). Comparison of pre- and post-treatment TAM values will be significant; *however, comparison as a percentage of normal value is invalid.*

TAM is a term applied to one finger, and is analagous to TPM in calculation except that only active motion is recorded, not passive.

1. Sum of active MCP flexion, + active PIP flexion, + active DIP flexion.

For example:

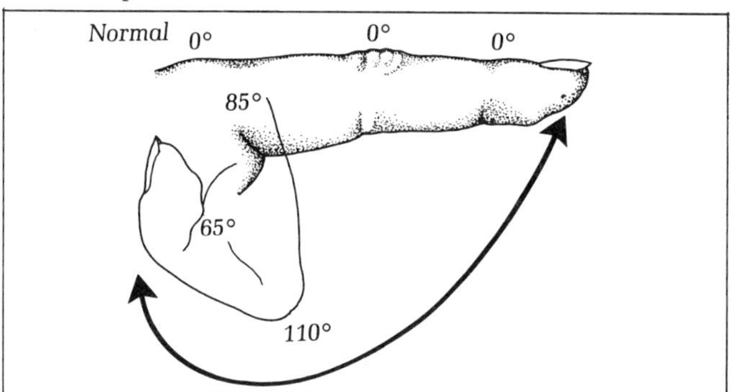

Figure 56
Normal

Active	Flexion	Extension Lack
MCP	85°	0°
PIP	110°	0°
DIP	65°	0°
Totals	260°	0°

Total Active Motion (TAM)
260° − 0° = 260°

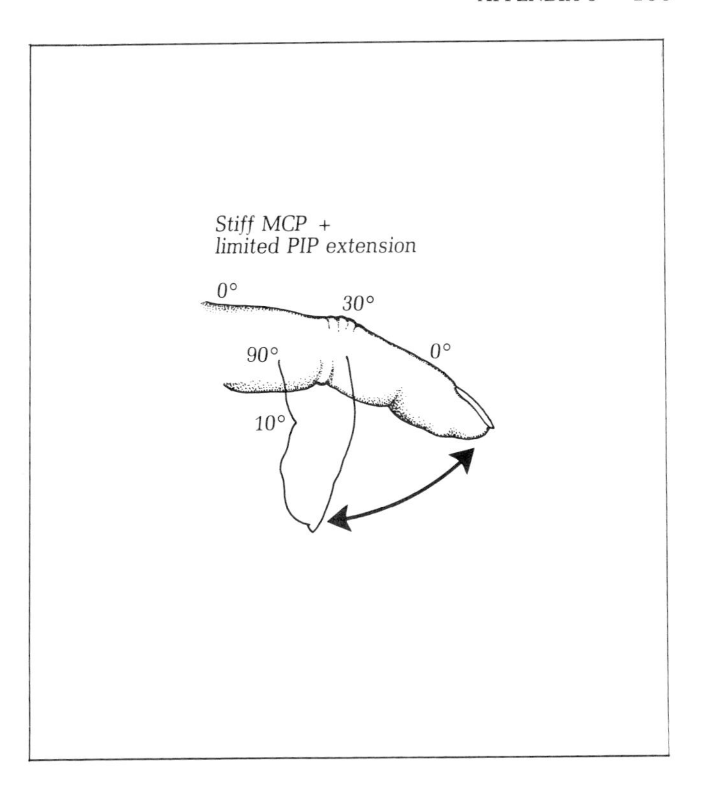

Figure 57

Stiff MCP + Limited PIP Extension

Active	Flexion	Extension Lack
MCP	0°	0°
PIP	90°	30°
DIP	10°	0°
Totals	100°	30°

Total Active Motion (TAM)
100° - 30° = 70°

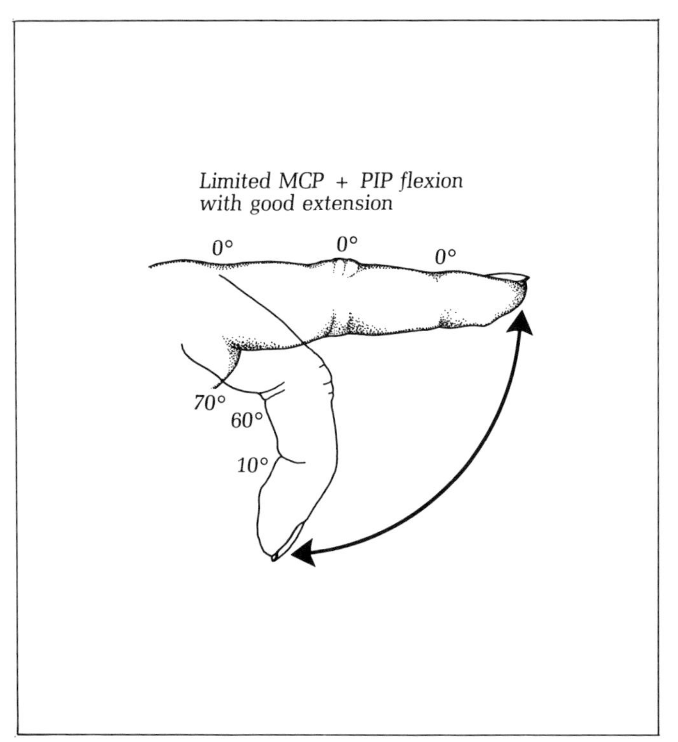

Figure 58

Limited MCP + PIP Flexion with Good Extension

Active	Flexion	Extension Lack
MCP	70°	0°
PIP	60°	0°
DIP	10°	0°
Totals	140°	0°

Total Active Motion (TAM)
140° - 0° = 140°

2. Minus sum of incomplete active extension (if any is present).

It is of critical importance to emphasize that this system of measuring and recording joint motions

1. is for a single digit
2. is to indicate the total motion of that digit in degrees
3. is to compare this to subsequent measurements of *that same digit* or the corresponding normal digit of the opposite hand in the *same* patient to determine if the patient is gaining or losing motion
4. absolutely is *not* intended to calculate a *percentage* of 'functional improvement or loss'
5. absolutely is not intended to calculate a 'percentage of impairment'.

Note that some finger joints are more important than others in digital function. Furthermore, note that 'function and impairment' involve many other factors as well, such as sensation.

VASCULAR STATUS

Patients who have vascular repair are evaluated in the following manner (not acutely, but late):

1. Examine for tissue survival.
2. Objective evidence of patent vessels by Allen test and/or ultrasonic pulse detector.
3. Revascularized part examined in resting and post-exercise state by one of several methods:
   a. presence of capillary filling.
   b. physiologic testing such as ultrasonic pulse detector, skin temperatures, etc.

   When possible, comparison with evaluation before and after three minute tourniquet ischemia.
4. Evaluation regarding cold tolerance of the part.

Ratings

1. Failure — no survival.
2. Poor — tissue survival.
3. Fair — objective evidence of patent vessels.
4. Good — function not limited by circulation.
5. Excellent — no cold intolerance.

Index

Abbreviations, 100
Adductor pollicis muscle, function test, 25
Amelia, 85
Anatomy, 4, 6, 102–105
Anomalies, congenital, 84–88
Aphalangia, 85
Arches of hand, 43
Arthritis, degenerative, 73, 74
  rheumatoid, 70, 72
Arthrogryposis, 87
Axial compression and rotation in degenerative arthritis, 73

Bennett's fracture, 58, 59
Bite infections, 99
Blood vessels clinical assessment after repair, 111
Bones, anatomy, 43
  fusion, congenital, 87
  terminology, 5, 6
  tumours, 92
Bouchard's nodes, 73
Boutonnière deformity, 66

Camptodactyly, 87
Cancer, 89
Carpal tunnel syndrome, 77, 79
Carpus, anatomy, 43
Circulation, evaluation, 41, 111
Claw hand, 66, 69
Clinodactyly, 87
Congenital anomalies, 84–88
Cyst, mucous, 89, 91

Deficiencies, congenital, 84
Deformities, acquired, 64–83
deQuervain's tenosynovitis, 73, 76
Differentiation, congenital failure, 86
Duplication of parts, congenital, 87
Dupuytren's contracture, 70, 71

Examination, physical, 7
  specific systems, 11–47
Extensor muscles, function tests, 16–24

Felon, 93, 94
Fingers, absence, congenital, 85
  deformities, acquired, 64–67, 70
Finkelstein's test for deQuervain's tenosynovitis, 73, 76

Flexor muscles, function tests, 12–16
tendon sheath, infection, 93, 94
Fractures, 53–61
Bennett's, 58, 59
boxer's, 58
intra-articular, 56
terminology, 54
Froment's sign, 25, 27

Ganglion, 89, 90
Gigantism, 88
Grip strength assessment, 7, 106
Guyon's canal, 32, 35

History, clinical, 3
Horseshoe infection of flexor tendon sheath, 95
Hypoplasia, 88
Hypothenar muscle function test, 29, 31

Infection, 93–99
Interosseous muscle function, 25, 28, 31
Intrinsic muscle tightness, 31

Joints, anatomy, 43
dislocations, 61
injuries, 56
terminology, 5, 6

Lacerations, examination, 49–52
Lobster claw hand, 86

Mallet finger, 64, 65
Median nerve, anatomy and function, 32
Melanoma, malignant, 89
subungual, 89
Metacarpals, anatomy, 43
Metacarpophalangeal joint, dislocation, 61
Motion, assessment, 7, 107–111
Muscles, 102–105
examination, 11
extrinsic, function testing, 12–24
function tests, 12–31
intrinsic, function testing, 25–31
Musculotendinous function, tests for, 16–24

Nerves, 31–41, 102–105
anatomical variation, 39

Overgrowth, 88

Paronychia, 93, 94
Phalen's sign in carpal tunnel syndrome, 82, 83
Phocomelia, 86
Pinch strength, assessment, 7, 107
Polydactyly, 87

Radial nerve, anatomy and function, 36
Rheumatoid arthritis, 70, 72

Scaphoid, fracture, 58, 60
Sensibility, assessment, 39, 106
Skin, examination, 11
Space infections, 95, 97
Strength, assessment, 7, 106
Surface anatomy, 6
Swan neck deformity, 66, 68
Syndactyly, 87
Synphalangism, 87

Tendon compartments, examination, 16
sheath, giant cell tumor, 89
infection, 93, 94
Tenosynovitis, 73
purulent, 93
Terminology for hand structures, 6
Thenar muscles, atrophy, 80, 83
function tests, 24
space, infection, 95, 97
Thumb, adduction test, 25
dislocation, 61
saddle joint, 45, 47
Tinel's sign in carpal tunnel syndrome, 81, 83
Trigger thumb and finger, 77, 78, 87
Tumors, 89–92

Ulnar collateral ligament, injury, 61
nerve, anatomy and function, 32, 34, 36
palsy, 69
Undergrowth, 88

Wrist, tendons, examination, 16